# ARCHITECTURE AND PROTOCOLS FOR HIGH-SPEED NETWORKS

# ARCHITECTURE AND PROTOCOLS FOR HIGH-SPEED NETWORKS

*EDITED BY*

**Otto Spaniol**
*Technical University of Aachen*
*Aachen, Germany*

**André Danthine**
*University of Liege*
*Liege, Belgium*

**Wolfgang Effelsberg**
*University of Mannheim*
*Mannheim, Germany*

KLUWER ACADEMIC PUBLISHERS
BOSTON / DORDRECHT / LONDON

Library of Congress Cataloging-in-Publication Data

Architecture and protocols for high-speed networks / edited by Otto
  Spaniol, André Danthine, Wolfgang Effelsberg.
      p.    cm.

    1. Computer network protocols.  2. Computer network architectures.
  3. Asynchronous transfer mode.  4. Computer networks.   I. Spaniol,
  Otto, 1945-   . II. Danthine, A.  III. Effelsberg, Wolfgang.
  TK5105.55.A73  1994
  004.6'5--dc20                                            94-31167
                                                              CIP

ISBN 978-1-4419-5148-9

Published by Kluwer Academic Publishers,
P.O. Box 17, 3300 AA Dordrecht, The Netherlands.

Kluwer Academic Publishers incorporates
the publishing programmes of
D. Reidel, Martinus Nijhoff, Dr W. Junk and MTP Press.

Sold and distributed in the U.S.A. and Canada
by Kluwer Academic Publishers,
101 Philip Drive, Norwell, MA 02061, U.S.A.

In all other countries, sold and distributed
by Kluwer Academic Publishers Group,
P.O. Box 322, 3300 AH Dordrecht, The Netherlands.

*Printed on acid-free paper*

# PREFACE

This book contains a selection of contributions (together with most recent additions and modifications made by the respective authors) which were presented at the First International Workshop on "Architecture and Protocols for High-Speed Networks", Schloss Dagstuhl, Germany (August 30 - September 3, 1993). The workshop was attended on an invitation basis by 35 international experts who discussed about actual problems and solutions in the rapidly expanding area of high-speed networking.

Major topics of the seminar were:
- switched networks, in particular ATM
- local and metropolitan area networks
- new network and transport layer protocols
- network applications, in particular multimedia applications
- protocol implementation on multiprocessors, and
- formal description techniques.

The general purpose of the workshop was to bring together telecommunications engineers and computer scientists, two groups of people who not very often have a chance to talk with each other. One of the hot topics was the status and future of ATM (Asynchronous Transfer Mode). Although ATM was initially designed to provide a wide-area high-speed telecommunications infrastructure, almost all of the installations and of the practical experience is concentrated on ATM switches in a local environment.

The new generation of applications in high-speed networks will contain multimedia data streams, i.e. digital audio and video. Continuous media streams, however, require transmission with guaranteed performance, in particular guaranteed bandwidth and bounds for delay and jitter. In addition to that, many multimedia applications will require peer-to-multipeer communication. Guaranteed performance can only be provided with resource reservation in the network, and efficient multipeer communication must be based on multicast support in the lower layers of the network.

Several manuscripts deal with internal structures for high-speed communication nodes. It is generally agreed that the performance bottleneck is currently in the end systems, upper layers and applications rather than in the

MAC adapters, on the links or in the switch fabrics. Parallel implementation of protocols on multiprocessors is considered as a promising solution.

All presentations were really excellent but due to the page number limitation the editors had to make a selection; less than 50 percent of the offered material could be included in this book. After a lot of discussions between the editors it was decided to concentrate on two areas which are in the center of interest for research and implementation of communication systems:
- protocol related aspects (switched networks, ATM, MAC layer, network and transport layer, traffic control, parallel processing,...)
- services and applications (multimedia systems, quality of service,...).
Even with such a concentration it turned out that a further selection was unavoidable. Since almost all authors delivered their manuscript in due time (the editors had never expected such an acceptance rate) the final versions had to be even more 'condensed' but this final procedure led to a significant increase in the quality of presentation (confer the following bon mot made by a famous person: "I'm writing a long letter since I couldn't afford the time to formulate a shorter one"; we apologize for writing a rather long preface!).

In most cases, it is difficult or impossible to exactly associate the manuscripts of the book to exactly one area since all relevant work must reflect aspects of different topics. The manuscript ordering, nevertheless, tries to follow basically the 'classical' rule: from lower layers to higher layers and finally to applications.

The manuscripts include a lot of new ideas resulting from very lively discussion rounds which were held in evening sessions during the workshop itself. The atmosphere of Dagstuhl castle was extremely positive for such intensive and fruitful discussions. Moreover, the fact that the participants were real experts in the field became a guarantee for critical but constructive comments; the editors are convinced that this interaction is visible in the manuscripts which have been updated several times, which have been thoroughly refereed and which present original unpublished material.

**Otto Spaniol**　　　　**Technical University of Aachen, Germany**
**André Danthine**　　　**University of Liège, Belgium**
**Wolfgang Effelsberg**　**University of Mannheim, Germany**

# CONTENTS

# 1

# SIZE AND SPEED INSENSITIVE DISTRIBUTED QUEUE NETWORK

Z.L. Budrikis, A. Cantoni, J.L. Hullett

*Australian Telecommunications Research Institute, Curtin University of Technology, Perth, Western Australia*

## ABSTRACT

A shared medium ATM switch in the form of a distributed queue dual bus (DQDB) network is described. Access to the network is in multiple stages, a separate stage for each branch of the network. Performance of the queue protocol is affected by distance within a stage, but not by distances between stages. In consequence a DQDT network can be of arbitrary extent without thereby limiting the rate, and DQDT networks can be designed with a speed-distance product that is orders of magnitude larger than is possible for DQDB, FDDI and other high speed LANs.

A DQDT network is asymmetrical in operation: Information is gathered from leaves towards the root of the concentrator tree, and is dispersed from the root towards the leaves of the distributor tree. The asymmetry makes DQDT a suitable component of a star-based network. It can serve as the first or concentrator stage in a wider ATM network.

## 1 INTRODUCTION

A shared medium network may be viewed as a time-shared communications server. As is well known, first-in-first-out, or FIFO queue service, outranks all other forms of service in efficiency and quality of performance. Little wonder then that, despite comparatively recent appearance, the distributed queue dual bus (DQDB) network has achieved prominence and attracted much attention and acceptance as a network standard [3,13,12,7].

As was pointed out by Mischa Schwarz [14], the performance of an M/G/1 queue and, for the case of constant length packets, of the M/D/1 queue, give an upper bound to the performance of *all possible* multiaccess strategies, where performance is judged by shortness of waiting time and smallness in the variance of the average waiting time. Strictly, the bound is established only for Poisson arrivals. But it appears true for all arrival patterns. Given a single server of fixed rate, the average waiting time will be minimum if the server is never idle while there is a customer waiting for service. This is true for all queued service, irrespective of service discipline. A further significant minimum can be attributed to the FIFO discipline, namely that a server, serving in FIFO order, achieves not just a minimum in the average waiting time, but also a minimum variance in the average waiting time. This far-reaching observation was made by Kingman [8].

The variance in the waiting time is an important attribute of a communications service, particularly when the communications include real-time signals, because in reconstituting the time fabric at the receiver, the *largest* delay that occurs during the course of a connection contributes directly to the end-to-end delay of the signal [10]. Generally, a smaller variance in the delay signifies also a smaller largest delay that is possible.

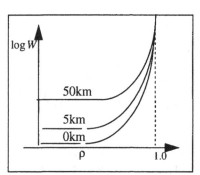

**Figure 1**   Waiting time with central scheduler

While a queue service gives the shortest possible waiting time, a *central* scheduler in a distributed network will not achieve this. The central scheduler would approach the theoretical bound only if both its processing time and the propagation time between it and the terminals on the network were negligible. However these delays are increasingly significant as the speed and size of the network grow. In fact the processing time and the two-way propagation time become an added constant to the waiting delay. This is illustrated in Figure 1, bringing out the effect of distance between terminal and scheduler.

Even with the additional delay due to signal propagation, the central scheduler will outperform most other practical multiaccess schemes. For instance, it will give a better performance in terms of maximum throughput, waiting time, and variance in waiting time than can be obtained with token passing, or with cyclic service, or with round-robin multiplexing. All these schemes have reduced performances due to larger protocol produced server idle times.

DQDB is exceptional among known practical schemes in performing better than the central scheduler. DQDB shares with the central scheduler FIFO queueing, but gives service that is almost totally free of protocol induced idleness and that does not suffer in the same way as that of the central scheduler from signal propagation delay. In DQDB access delay, averaged over all access points, is very little larger than the average waiting time in an ideal FIFO queue, and the maximum throughput remains essentially 100 per cent even for very large networks.

Aspects of performance on which DQDB does deteriorate significantly with increasing network size and speed are adherence to strict FIFO service and to equality of service to all terminals under heavy load. Even though non-adherence to FIFO only affects the variance in waiting delay, and unfairness under overload can be effectively be eliminated at small cost in total throughput by bandwidth balancing [6], there is an effective limit on the practically possible size of a DQDB network. At 155 Mbps and cell size of 53 octets, the maximum end-to-end length of network may be put at 40 km [2].

We report the invention of another distributed queue network, based on a dual tree rather than dual bus topology, or DQDT [1], which in effect can circumvent the distance dependent problems of DQDB. It does so by employing staged queueing, thereby introducing the ability to restrict the distance over which the distributed queue protocol is implemented, without restricting the overall distance spanned by the network.

## 2 DESCRIPTION OF THE DQDT NETWORK

Just like DQDB, the DQDT is a network with oppositely directed unidirectional information flows, with the information in fixed size, contiguous time slots or cells. However, unlike DQDB which is operated symmetrically, DQDT is asymmetrical. Information is written only on the one flow stream, the *concentrator tree*, and read from the other, the *distributor tree*. The buses of DQDT are referred to as trees because, as illustrated in the schematic of Figure 2, they may be in the form of trees, with branchesup to arbitrary order. The network is a *dual tree* because the concentrator and distributor are each a tree, one with information flow towards its root, the other with flow away from root. In the case of the network functioning as a stand-alone switch, the information that arrives at the concentrator root would be transferred directly to the distributor root, as indicated in Figure 2. The network may also function as a first stage concentrator/distributor element in a larger switched network when the roots would be connected to input and output of a switch port.

The disk-shaped elements in Figure 2 are media adaptors (MAs), and the square elements represent terminal equipments (TEs). As a general rule, MAs are three-port devices with two in-line ports and one lateral port. Tree limbs (stem or branches) are formed by in-line interconnection of MAs. The lateral port of an MA can take the attachment of either a TE or of a next higher order branch of the tree.

In ordinary LAN practice the media adaption function would be contained in the terminal, without any exposed interface between MA and TE. In DQDT an exposed interface is mandatory because, for one and as already noted, a branch is created by plugging into a media adaptor a section of network in place of a terminal, and for another and more importantly, to exploit the distance capabilities of DQDT to the full, media adaptors on the same branch should be as close to each other as possible, suggesting that they be clustered into a hub, and hence that terminals and the media adaptors to which they attach, be physically separated from each other.

The function of the whole DQDT network becomes clear from the function of a single media adaptor. Figure 3 shows the block schematic of one media adaptor.U and V are shown as plugs and are at the crown end of the adaptor, V on the concentrator and U on the distributor. X and Y are sockets at the root end, and A

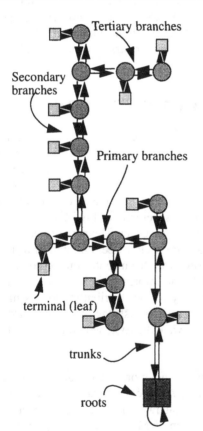

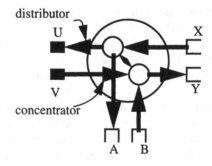

**Figure 2**   The DQDT Network          **Figure 3**   Schematic of a media adaptor

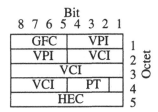

Bit
8 7 6 5 4 3 2 1

| GFC | VPI | 1 |
| VPI | VCI | 2 |
| VCI | | 3 |
| VCI | PT | 4 |
| HEC | | 5 |

Octet

**Figure 4**  Header format

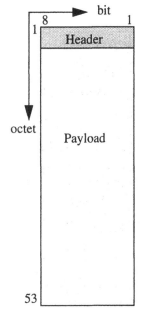

**Figure 5**  Cell format

and B are lateral sockets. Information enters the MA from a TE, or equivalently from a higher order branch, through B and has to be inserted into the concentrator stream which enters at V and exits at Y. Similarly, information enters in a distributor stream through X and is passed on to U as well as copied to terminal, or branch, through A. To insert the information into the concentrator stream, the MA uses the distributed queueing protocol.

Distributed queueing requires twin information streams, "upstream" and "downstream", carried in slots and accessed by all participating nodes. The information format must include an access control field. There are no further restrictions. Many different slot formats are possible. For the sake of description, and with an eye on future application for our network, we take the format defined for ATM at the user-network interface (UNI) in B_ISDN [4]. The slot, or cell as known in B_ISDN, is of 53 octets of which five octets are the header and 48 octets payload, as shown in Figure 4. The header format, as defined at the UNI, is shown in Figure 5. The very first four bits make up the Generic Flow Control Field (GFC) which can serve as access control information field for the distributed queueing in DQDT.

GFC protocol and procedures are still in the process of definition in ITU, with finality expected in 1995. The GFC that is being defined in ITU, is for control across the S and T reference points in the ITU defined customer premises network. and therefore will apply to the lateral, or TE-to-MA, flow in DQDT. The definition will provide for admission control on ATM connections that have no guaranteed bandwidths (the class of 'controlled' connections), and will give an effective back-pressure control through a credit reset/no-reset scheme.

The queue access control along branches and trunk must provide control for all access, including on connections that have guaranteed bandwidths.To achieve this it can be similar to the Access Control Field (ACF) of IEEE 802.6 in providing for queues of multiple priorities. In queueing, a node sends requests for empty slots to nodes upstream of itself, which for DQDT would mean that requests are sent on the distributor tree. There is no essential role for GFC on the concentrator where it could for instance be defined to indicate whether the particular cell is on a bandwidth resourced or unresourced connection.

We propose that the GFC on the concentrator be $(S, R_0, R_1, R_2)$ where S is a STOP bit, $R_0$, $R_1$, and $R_2$ are REQuest bits, respectively at priority levels 0, 1, and 2. Priority level 0 is the lowest and is used for unresourced connections corresponding to the 'controlled' connections of the ITU definition, while Priority levels 1 and 2 are for resourced connections. The highest level would be intended for connections that go over the T reference point to the public network, and the middle level for intra-premises or local connections. The STOP bit, when set, would stop access for one cell period to priority level 0 cells, i.e. to cells on 'controlled' connections.

Distributed queueing is implemented separately on each limb (trunk or branch) of DQDT. A higher order branch offers traffic to a lower order branch (or trunk) that it terminates onto, no differently from the traffic offered by a terminal. Thus at each priority level, the concentrator of DQDT a cascade of queue stages, as illustrated by Figure 6. Assuming that the distributed queueing performs ideally, each stage is equivalent to a FIFO buffer with parallel inputs.

The servers are in all cases slotted and have different effective rates for the three levels of priority: At priority level 2 the server is at bus rate, at level 1 it is at bus rate less the actual service at level 2, and at level 0 it is at bus rate less the actual service at levels 2 and 1. Queues at priority levels 2 and 1 are stable because inputs are regulated by contracts at Call Admission, and at priority level 0 are made stable by back-pressure control exerted through the GFC.

A distributed queue will approximate very closely the ideal single-server queue and will adhere strictly to priorities in the case of priority queueing, as long as REQuest propagation delays are shorter than one cell period [11]. With the queueing independent on each branch, the relevant delay is confined to a branch and, more precisely, to the node-to-node delay along the distributor member of the branch. Significantly, any delay between branch point and the nodes has no relevance to the queue protocol and hence a branch may have an arbitrary length from branch-point to first node, without affecting the functioning of the distributed queue protocol. To assure then ideal single-server queue behaviour and strict adherence to priority in service, the number of nodes on a branch and their spanned distance must be so limited that the maximum protocol delay does not exceed the set ceiling of one cell period.

The limit on spanned distance can be met easily by placing all nodes of a branch into a single cluster or hub. Figure 7 shows (with artistic license) how a DQDT network, with only one cluster per branch, might look. A cluster may have the extent of just a backplane and have a physical propagation span of less than one

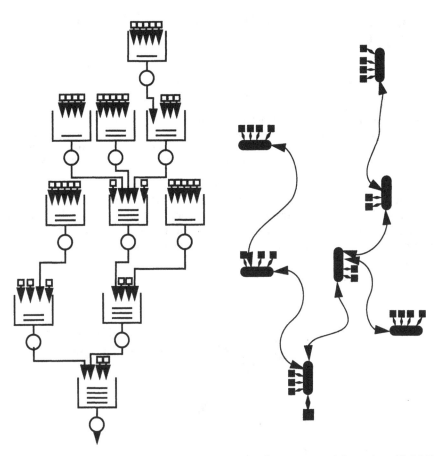

**Figure 6** Staged queueing in DQDT concentrator

**Figure 7** Example of cluster-based DQDT network

metre. The propagation delay would then at most be 3 ns. Assuming 53 octet cells, a cell period of 3 ns corresponds to a bit rate of 140 Gbps.

However a generally larger delay than due to physical propagation should be expected from latency through media adaptors. This will depend on implementation and should be in number of bits, and nearly independent of rate of network. Therefore, rather than limiting the distance, it would limit the number of media adaptors that can be accommodated on a single branch. A practical limit might be, say, 12 MAs per cluster. This, of course, need not be the limit on terminals supported by a cluster since MAs with multiple (lateral) ports are possible.

# 3  DQDT FIRST STAGE IN SWITCHED ATM NETWORK.

The DQDT is exceptionally capable in terms of distance that it can span, and is outstanding in the quality of shared service that it can provide. In common with all shared-medium networks, it will support broadcasting and multicasting. It has a very simple access protocol. If one was interested in a LAN that span the wide area, DQDT would be a very suitable candidate. It can be a stand-alone network by connecting the concentrator tree root to the root of the distributor tree, as was indicated in Figure 2. As any other LANs, DQDTs could be bridged, with half-bridges taking the positions of terminals on the networks. But the DQDT network is even more interesting as a component in a wider ATM network than as an ultra high speed or wide area LAN.

As already noted, DQDT is asymmetrical in character that gives it a star orientation: The roots of its concentrator and distributor are ready terminals by which it can be connected to a further switching stage of a wider star network. Figure 8 illustrates the case of the smallest possible wider network, namely where the next switching stage is a 2x2 switch and has its second pair of terminals connected to the public network at the T reference point.

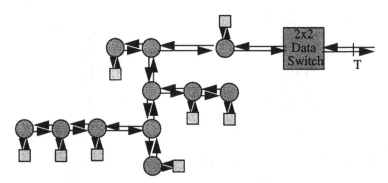

**Figure 8**   DQDT as concentrator/distributor in switched ATM network

The 2x2 switch is not a common ATM switch, but rather an ATM level DataSwitch [5]. A data switch differs from an ordinary cell switch in that it can also perform statistical multiplexing. For instance, it would multiplex traffic from any number of virtual circuits, individually unconstrained in bandwidth, onto a single, bandwidth-limited virtual path. It does this by virtue of a large buffer and judicious cell discard when in congestion, the discard being of data frames rather than unconnected cells. It also implements priority: Higher priority cells are served ahead of lower priority cells.

The system of Figure 8 provides LAN services locally on the DQDT. Terminals attached to the DQDT can communicate with each other on a one-to-one and on one-to-many bases over unresourced or 'controlled' connections. Access for their traffic is regulated through the GFC. The terminals can also communicate locally as well as remotely on resourced or 'uncontrolled' connections. Their bandwidths are guaranteed by the extended queue protocol [9] in conjunction with the DataSwitch. The terminals can also communicate in LAN manner with remote sites over the public network, assuming that the public network provides virtual path interconnections onto which the DataSwitch can multiplex individually unresourced remote traffic

## 4 SUMMARY

A novel distributed network, the Distributed Queue Dual Tree (DQDT) has been described. The network has outstanding distance and speed capabilities. It derives its distance capability by giving queued access in multiple cascaded stages of queueing, without requiring any restriction on the physical distance between queueing stages.

The network is asymmetrical in character, having a concentrator tree that channels all information from leaves towards the root and a distributor tree that broadcasts information from root to leaves. The two roots are natural terminal points for connecting the network to the next higher stage in a star-based network. So connected, the DQDT network will act as a concentrator/distributor first stage of the wider network.

Such connection and role can be seen as being particularly attractive in the context of ATM and B_ISDN. The DQDT first stage would give to the ATM network LAN service qualities, without compromising any of its circuit-switching capabilities

## REFERENCES

[1] Patent applied: Z.L. Budrikis, A. Cantoni, and J.L. Hullett, Distributed Queue Dual Tree Digital Network, AIPO Provisional Specification Lodgement PL 9705, 1993

[2] Z.L. Budrikis, G. Mercankosk, M. Blasikiewicz, M. Zukerman, L. Yao and P. Potter, A Generic Flow Control for B_ISDN, Australian Telecommunications Research, Vol 26, 1992, 55-65.

[3] Z.L. Budrikis, J. L. Hullett, RM Newman, D. Economou, F. M. Fozdar, and R. D. Jefferey, QPSX: A Queued Packet and Synchronous Circuit Exchange, Proc 8th ICCC (North-Holland, Amsterdam, 1986) 288-293.

[4]  TU CCITT Recommendation I.150, B_ISDN Asynchronously Transfer Mode Functional Characteristics, Geneva, 1991; ITU CCITT Recommendation I.361, B_ISDN ATM Layer Specification, Geneva, 1991

[5]  Curtin University of Technology, Data switch for asynchronous mode-based networks, Provisional Patent Specification, PL 7981, AIPO, March, 1993

[6]  E.L. Hahne, A.K. Choudhury and N.F. Maxemchuck, Improving the Fairness of Distributed-Queue-Dual-Bus Networks, Proc IEEE INFOCOM '90, San Francisco, 175-184.

[7]  IEEE Standards for Local and Metropolitan Area Networks: Distributed Queue Dual Bus (DQDB) Subnetwork of a Metropolitan Area Network (MAN), 802.6-1990

[8]  J..F.C. Kingman, The Effect of Queue Discipline on Waiting Time Variance, Proc Camb Phil Soc, Vol 58, 1962, 163-164

[9]  G.Mercankosk, Z.L. Budrikis, A. Cantoni, Extended Distributed Queueing, Network Research Laboratory, ATRI, NRL-TM-23, June, 1992.

[10] Guven Mercankosk and Antonio Cantoni, Characterization of a CBR Connection over a Channel with known Bounded Delay Variation, Proc IEEE Infocom '93, San Francisco, March 1993, 1170-1177

[11] R.M. Newman, The Distributed Queueing Protocol for an Integrated Communications Switch, University of Western Australia, Ph. D. Thesis, July 1987

[12] R. M. Newman, Z. L. Budrikis and J. L. Hullett, The QPSX MAN, IEEE Communications Magazine, Vol 26, 20-28, April, 1988

[13] R. M. Newman and J. L. Hullett, Distributed Queueing: A Fast and Efficient Packet Access Protocol for QPSX, ibid, 294-299

[14] Mischa Schwarz, Telecommunication Networks, Addison-Wesley Publishing Company, Reading, Massachusetts, 1987, 460-461

# 2

# HIGH PERFORMANCE ACCESS MECHANISMS FOR SLOTTED RINGS

## S. Breuer, T. Meuser, O. Spaniol

*Computer Science Department, Aachen University of Technology*
*D-52056 Aachen, Germany*

## ABSTRACT

The slotted ring enhanced by spatial slot reuse is one potential approach for data transfer and admission control in high performance networks. This paper presents a method for calculating the possible gain in transmission capacity by slot reuse which depends on the given traffic distribution. Using these results we discuss different media access control schemes implemented in several protocols (ATMRing, CRMA-II, and MetaRing) with respect to their capability of maintaining fairness and bounded access delays.

## 1 INTRODUCTION

Rapid progress in the evolution of optical fiber transmission has resulted in increasing channel capacities for high speed communication systems. Future networks will have raw data transmission capacities of multiple Gbit/s. Such rates enforce a redesign of media access control (MAC) protocols to minimise the tradeoff between growing bandwidth and the fixed propagation delay [1]. For example, the FDDI timed token protocol becomes inefficient for increased data rates and network sizes [2].

Designing Gbit/s networks is much more than just pushing bits over some physical media at high speed [3]. Acceptable qualities of service have to be guaranteed to provide the end-users with the benefits of high performance networking. Provision of a broad range of services is essential for backbone networks, high speed computer environments, visualization, or distributed multimedia. Today's LAN and MAN protocols use hybrid schemes to support delay-sensitive and bandwidth-consuming applications. To enable interconnection to the emerging ATM technology an integration of isochronous and non-isochronous data through a common asynchronous service is crucial [4].

Future MAC protocols will have to be scalable with respect to geographical size, transfer rate, and number of nodes. They have to provide high efficiency in bandwidth utilization and bounded access delays in a multi-service environment [5]. Furthermore, interconnection to ATM-based systems is essential.

For shared-medium LANs and MANs the slotted ring architecture with spatial slot reuse is favoured as transmission system. It allows a significant gain in performance by providing network capacities far beyond the nominal data rate. Unfortunately, this basic mechanism is unfair, and without an additional fairness control some stations may suffer from severe performance degradation. This problem has been addressed by different media access control protocols such as **ATMR** (ATMRing) [6,7], **CRMA-II** (Cyclic Reservation Multiple Access) [8,9], and **MetaRing** [10]. These protocols have been designed for networks operating as Gbit/s LANs as well as for access control for interfacing users to ATM systems by a shared medium. They operate on the dual counter rotating ring topology, employ spatial slot reuse with shortest path routing, and allow immediate access to the network under low and medium load.

Several performance studies of these protocols have been published, cf. [11-16]. The presented manuscript compares their performance for both saturated and non-saturated senders. The basic protocol concepts are summarised, and a new mathematical approach is introduced to calculate the maximum capacity of a ring network which employs slot reuse as a function of the given traffic pattern. Using these upper bounds we evaluate the simulation results for the MAC protocols.

# 2  BASIC ACCESS SCHEMES FOR SLOTTED RING NETWORKS

This section gives a brief description of the mechanisms used by ATMR, CRMA-II, and MetaRing, respectively. The discussion is restricted to the dual counter rotating ring topology with destination release where stations are allowed to access free or only just released slots immediately. No priority mechanisms are considered.

In order to provide bounded delays and fairness as well as to prevent 'starvation' of stations which never receive any free slot, there are two basic access control concepts. CRMA-II uses dynamic unfairness detection with explicit backpressure. An alternative approach is used by ATMR and MetaRing: the negotiation of periodically renewed transmission quota for every station.

## 2.1   ATMR

The ATMRing provides fairness control by means of a cyclic reset scheme and a distributed window mechanism with many similarities to the Orwell protocol [17]. The mechanism is based on a monitoring system: at every station an access unit (AU) observes which station was the last active sender. If the last active AU detects inactivity of all other AUs, it generates a reset immediately after its own transmission. Inactivity monitoring is performed by every active AU overwriting a *busyaddress* field in the header of every cell with its own address. Thus, every AU receiving a slot with its own busyaddress assumes that all other AUs are inactive.

The reset causes an AU to set up its *window counter* to a negotiated window size. This counter is decreased each time the AU uses a free slot. As the window counter expires, the AU is forced into inactive state, where it is not allowed to send any data until reactivation by the next reset. Thus, it is guaranteed that every station uses a maximum number of cells between two consecutive resets. A station also becomes inactive if it has no more data to send. It is reactivated by arriving data at the transmit queue.

## 2.2   CRMA-II

The CRMA-II ring provides immediate access to free/gratis slots for every non-blocked station. Every access to a free slot increments the *transmit counter* of the station. Traffic regulation is performed by bandwidth scheduling in successive transmission cycles.

At the beginning of every cycle a *scheduler* sends a reserve command to ask all stations for intended sending requests and the actual value of their transmit counter. Based on this information the scheduler calculates a fairness threshold which determines numbers of request confirmations and deferments for every station. This is done in such a way that the sum of cells to be reserved and deferments becomes minimal in order to achieve minimal transmission throttling. Finally, the scheduler issues a confirm command which circulates around the ring informing every station about the new threshold.

Receipt of the confirm command causes a station to decrement its transmit counter by the given threshold, to actualise its state, and to initialise its confirm or deferment counter. If the actual value of a station's transmit counter is higher than the calculated threshold, it has to change from reserve mode to defer mode. In this case it gets no confirmations and has to pass as many free/gratis slots as its transmit counter exceeds the calculated threshold. Stations in reserve state are allowed to use every free/gratis slot. As long as its confirm counter is positive free/reserved slots can also be used. Immediately after sending the confirm command the scheduler marks all passing gratis slots as reserved until the number of allocated slots is reached. Subsequently, next cycle is started by sending a new reserve command.

## 2.3  MetaRing

In this proposal fairness and non-starvation are guaranteed using a continuously circulating hardware control message, called SAT-signal, which periodically renews the transmission quota of every station. This signal has preemptive resume priority, i.e., it can be inserted into the data flow at any time.

The network is initialised by negotiating Min- and Max-quota for every station. The quota represent the minimum and the maximum numbers of slots a station is allowed to send during one SAT-cycle. A station can use any free slot as long as it has not reached its Max-quota. A station is 'SATisfied' if it has no more data to send or if it has reached its Min-quota. A satisfied station immediately sends the SAT-signal to its neighbouring node. Otherwise, it keeps the signal until it becomes satisfied. When sending the signal to the neighbouring station, both quota are reset. This mechanism might be considered as being as a special version of a token scheme.

## 3  INVESTIGATED NETWORK SCENARIOS

The basic network scenario consists of 10 stations located at regular intervals on a dual counter rotating ring topology. Data rate of each ring is set to 1 Gbit/s, and a ring length of 10 km is assumed. The slot size is equivalent to that of an ATM cell.

The performance evaluation includes saturated traffic scenarios as well as variably loaded ones. With saturated scenarios, the basic behaviour of the protocols can be studied in an abstract way, revealing all their shortcomings. Variably loaded traffic scenarios reflect more realistic environments, since communication systems usually operate under light to medium load. In both traffic scenarios the given packet lengths vary according to a truncated hyperexponential (H2) distribution with coefficient of variation CV=2. The mean and maximum lengths are 700 and 4500 Bytes, respectively.

The traffic flow is characterised by symmetric or asymmetric destination matrices according to equation (2.1), which has been derived from a similar rule given in [16]. The main characteristics of this distribution is that packets are most likely directed to nearby nodes as specified by *locality parameter* p. Using this value parameters $p'_i$ are calculated selecting the *degree of locality* for every node in the range (0,1). Assuming N nodes, node i sends a packet to node j with probability

$$
P_{ij} = \begin{cases} \dfrac{p'_i(1-p'_i)^{20/N|i-j|-1}}{A_i} \,, & i \neq j, \\[2mm] 0 \,, & i = j \end{cases}
\tag{2.1}
$$

$$
with \ \ A_i = \sum_{j=1}^{N} p'_i(1-p'_i)^{20/N|i-j|-1} \quad and \quad p'_i = p^{10/N|N-i|+1}
$$

This equation allows to determine an individual degree of locality for every node and to change asymmetry of the destination matrix of the network. The factors $A_i$ normalise the distribution probability, and the quotients 10/N and 20/N preserve the characteristics of the traffic distribution also for an increasing number of stations. Using equation (2.1) we obtain the following traffic distribution: For p = 0,999 every node sends to its neighbours only; p = 0,001 leads to a uniform traffic distribution. Figure 1 illustrates the traffic distribution for p = 0,5. It shows that stations in the 'middle' of the ring transmit to destinations with a lower locality, because parameter $p'_i$ is influenced by the distance to station N. This leads to higher quota of slot receptions for stations located close to station N.

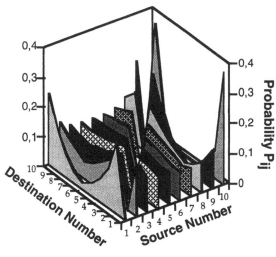

**Figure 1**   Traffic distribution for p = 0,5 and N = 10

The following metrics are used in our performance evaluation: *Network Throughput* is defined as the ratio of used slots to slots passed during a simulation run. *User Throughput* describes the amount of data which has actually been transmitted. *End-to-End delay* is the sum of the *waiting time* (which is the time between queuing a packet in the transmit buffer and the start of its transmission), the transmit time for the complete packet, and the propagation delay.

Performance evaluation is based on simulations using the ATLAS tool (**A**nalysis **T**ool for **L**ocal **A**rea Network **S**imulation) which has been developed at *Aachen University of Technology* [18]. It provides features for definition and evaluation of queueing networks and is based on the discrete event driven simulation technique. In order to achieve statistical significance, the duration of the simulation runs has been chosen according to keep the maximal ranges of the confidence intervals (of 99% based on the batch means test) within 5 % of the mean values.

# 4 THE PIPE MODEL

In this section, we introduce a model for calculating network capacity as a function of the traffic distribution matrix. We obtain results for both, uncontrolled networks where every station uses every available slot and networks maintaining global fairness for all active stations. The latter scheme gives the maximum throughput which can be theoretically achieved by an optimal algorithm having overall knowledge about past and future traffic flows within the network.

The pipe model is based on a single ring topology interconnecting N stations (0 to N-1). The link between station (i-1) and station i is called *pipe* $\Pi_{[i-1,i]}$ $(0 \leq i \leq N-1)$. Station change from *active* to *inactive* state if they do not send any data on the ring. Thus the *station set* K of all stations is divided into a set of inactive stations I and a set of active stations A $(K = A \cup I)$. In an activity vector $A = (a_i)_N$ an active station is marked by $a_i = 1$ and an inactive station is marked by $a_i = 0$.

Throughput is measured by the transmission rate normalised to the capacity of the medium. *Network throughput* $C$ is given by $C = \sum_{i \in \mathcal{K}} x_i$, where $x_i$ $(0 \leq i \leq N-1)$ is the transmission rate of station i. $x_i$ is split into partial rates, one for each destination according to the given *distribution matrix* $P = (p_{ij})_{N,N}$.

## 4.1 General pipe conditions

The major difficulty in throughput analysis is to capture the effect of varying network capacity due to changing occupations of the pipes.

**Definition:** (pipe throughput)
*Throughput $b_i$ of a pipe $\Pi_{[i-1,i]}$ ($0 \leq i \leq N-1$) is the sum of the partial rates of all stations $j \in K$, which are located before $\Pi_{[i-1,i]}$ to respective destinations $k \in I(i, j)$ behind $\Pi_{[i-1,i]}$ (see figure 2).*

Let
$$b_i = \sum_{j \in \mathcal{K}} \sum_{k \in I(i,j)} p_{jk} * x_j \quad \text{and} \quad \mathbf{B} = (b_0, \ldots, b_{N-1})$$

where:
$$I(i, j) = \begin{cases} \{k \in \mathcal{K} \mid i \leq k \leq j\}, & \text{if } i \leq j \\ \mathcal{K} - \{k \in \mathcal{K} \mid i < k < j\}, & \text{if } i > j \end{cases} \qquad (2.2)$$

∎

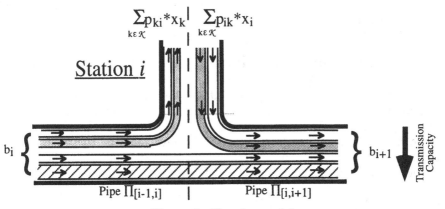

**Figure 2**   The pipe model

**Definition:** (pipe matrix)
Station $j \in K$ contributes to the traffic of pipe $\Pi_{[i-1,i]}$ ($0 \leq i \leq N-1$) with a rate $r_{ij} * x_j$ . The *pipe matrix* $\mathbf{R} = (r_{ij})_{N,N}$ is defined as

$$r_{ij} = \sum_{k \in I(i,j)} p_{jk} \quad . \qquad (2.3)$$

Using equations (2.2) and (2.3) the pipe throughput vector **B** is given by:
$$\mathbf{B} = \mathbf{R} * \mathbf{X} \qquad \text{where } \mathbf{X} = (x_0, \ldots, x_{N-1}) \qquad (2.4)$$

The station throughput vector **X** is called *valid* with regard to pipe matrix **R** if, for all $i$, $\sum_{j \in \mathcal{K}} r_{ij} * x_j \leq 1$ (i.e. no pipe capacity is exceeded).

∎

## 4.2 The uncontrolled scenario

First, we describe how the pipe model can be used to maximise the throughput vector $X$ in saturated scenarios. In the uncontrolled case every active station directly uses all slots addressed to it since there is no traffic regulation. Inactive stations pass their bandwidth to the downstream neighbour by so called *pseudo data streams*. Thus, in a *modified distribution matrix P'* an inactive station $i \in I$ is given by $p'_{i,i+1} = 1$ and $p'_{ij} = 0$, $j \neq i+1$. Replacing $p_{ij}$ by $p'_{ij}$ leads to a modified pipe matrix $R'$ in equation (2.3). Due to the saturated senders and the pseudo data streams the capacity of each pipe is completely utilized. Therefore, we obtain a set of linear equations for the throughput vector $X'$ :

$$R' * X' = \vec{1} \tag{2.5}$$

The dimension of the solution for these equations depends on the range of the matrix $(R', \vec{1})$ which is determined by the structure of the distribution matrices $P$ and $P'$, respectively. Finally, $X$ is given by $x_i = x'_i * a_i$ for all stations $i \in K$ .

## 4.3 Optimisation of a fairness threshold

Stations may be assigned different bandwidth in uncontrolled scenarios. In our definition, fairness is the behaviour to be applied to those stations which are competing for a common bandwidth. Thus, every saturated station should have the same chance to send. To achieve fairness, transmission by preferred stations has to be deferred. We now calculate the maximum network throughput in the case of global fairness.

**Definition:** (maximum fairness of saturated stations)
A scenario given by destination matrix $P$ and throughput vector $X$ is called *fair with throughput f* if
  • $X$ is valid with regard to $P$ and
  • $X = f * A$, i.e. all active stations get the same throughput f.
The *maximum fairness* $f_{max}$ is given by:
$$\|R * (f_{max} * A)\|_\infty = 1 \qquad\blacksquare$$

If throughput is maximal at least one pipe $\prod_{[i-1,i]}$ ($0 \leq i \leq N-1$) will become full, i.e. a bottleneck. Thus,

$$1 = b_i = \sum_{j \in \mathcal{K}} \sum_{k \in I(i,j)} p_{jk} * f_{max} \quad and \quad f_{max} = \frac{1}{\displaystyle\sum_{j \in \mathcal{K}} \sum_{k \in I(i,j)} p_{jk}} \tag{2.6}$$

To validate the analytic results by simulations, we consider several network configurations. Figure 3 presents numerical results for the uncontrolled scenario and for fair bandwidth sharing, denoted by OPT, in a network with 10 stations. As in this example, the calculated results correspond with the simulation results in all cases.

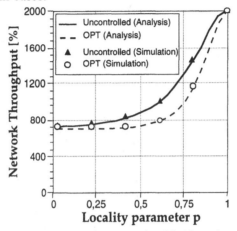

**Figure 3**   Throughput of a network with 10 stations

It is obvious that ring capacity grows with increasing locality of the traffic distribution given by locality parameter p. This parameter also determines the asymmetry of the traffic flow. Thus, capacity loss due to fair bandwidth sharing becomes higher until symmetry is reestablished if every station sends only to its neighbours.

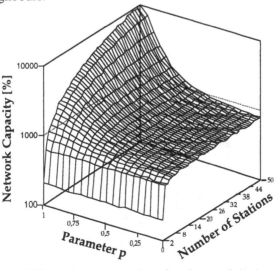

**Figure 4**   Network Throughput vs. number of stations and the locality parameter

A more general view of the capacity of uncontrolled networks in relation to the station number and the locality of the traffic distribution is given in figure 4. As mentioned before, the theoretical throughput obtainable in slotted ring networks depends on the degree of exploitation of spatial slot reuse. Setting p close to one results in an increased network throughput, up to N times of the medium capacity on each ring, since every station sends to its direct neighbour only. For uniform traffic distributions (i.e. small values of p) capacity of a <u>single ring</u> network is roughly doubled since the average distance between source and destination is half the ring length. On a <u>dual counter rotating ring</u> topology, shortest path routing reduces the average path of filled slots to a fourth of the ring length. Thus throughput of eight times the medium data rate (four times for each ring) can be achieved. It should be noted that these throughput values are only obtained for a large number of stations since the average distance between source and destination is given by (N+1)/4 (instead of N/4).

Calculation of an optimal fairness threshold is important for the performance evaluation of protocols which provide traffic regulation in slotted rings. Fairness will result in reduced throughput, and the only way to regulate the traffic flow is to force favoured stations to pass free slots to the disadvantaged ones. The best algorithm, called OPT, will exactly pass as many free slots as are necessary to compensate the unfairness. The capacity loss of the proposed protocols compared to OPT's throughput indicates the quality of the used mechanisms.

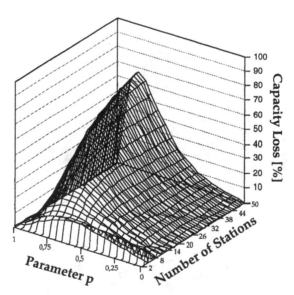

**Figure 5**  Capacity loss due to fairness

Throughput reduction for OPT due to fairness is shown in figure 5. Optimal traffic regulation does not affect the capacity for uniform traffic distribution where fairness is provided for all stations anywhere. With higher asymmetry of the traffic flow capacity loss increases substantially as compared with uncontrolled networks. The larger the asymmetry the more unused slots have to be passed to disadvantaged stations.

## 4.4   Model for non-saturated scenarios

We now extend the mathematical model to include throughput analysis in non-saturated scenarios where data is generated by an arbitrary arrival rate. The available ring capacity is divided into several parallel links which are saturated by different sets of active stations. These sets change according to the arrival rates and to the provided service rates for every station. The resulting saturated subscenarios are iteratively processed according to the algorithm for uncontrolled scenarios presented in section 4.2. Starting with an activity vector covering all stations, a throughput vector $X´$ is calculated according to equation (2.5) in each iteration. Subsequently, the station with the minimum quotient $q_i$ of arrival rate $d_i$ and service rate $x_i$ is separated. If the minimum $q_i´ < 1$, the service rate of the station met its arrival rate and the station became idle. Validity of the results is restricted to that part of the network capacity allocated under saturated conditions for which the algorithm has been defined. This portion is described by the minimal quotient $q_i´$ and determines the actual subscenario. In each subscenario only these parts of the resulting throughput vector $X´$ and the used ring capacity which results from the multiplication with $q_i´$ have to be considered. For the next iteration the activity vector has to be changed by marking all stations which became idle as inactive. This procedure continoues until either all stations became inactive or the minimum $q_i´ \geq 1$. In the latter case, the algorithm leads to a saturated subscenario which can be calculated completely. Total throughput of every station is given by the sum of partial throughputs of each subscenario.

Adapting the model for the calculation of the maximum network throughput in case of global fairness can be done by replacing the calculation of the throughput vector $X´$ by equation (2.6) for the upper fairness threshold in each iteration. The minimal quotient $q_i´$ is determined by dividing the arrival rate $d_i$ by the fairness threshold $f´$ rather than by service rate $x_i$. Since not all pipes are fully occupied in the optimal case, the remaining ring capacity is only reduced by the partial throughputs.

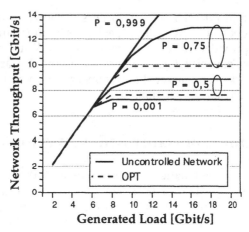

**Figure 6**  Network Throughput under increasing load conditions

Numerical results for the uncontrolled scenario and for fair bandwidth sharing versus the offered load on a ring with 10 stations are presented in figure 6. Throughput of the uncontrolled network compared to that of OPT shows that an optimised traffic regulation does not cause any capacity loss under low load or symmetric traffic distribution. An asymmetric traffic distribution results in some capacity loss to achieve fair bandwidth sharing. These optimal values will be used in the next section as upper bounds in the discussion of protocol efficiency. It should be noted that for non-saturated traffic scenarios the calculated results correspond with the simulation results as well.

# 5   COMPARISON OF TRAFFIC REGULATION SCHEMES

Inherent features of the slotted ring approach are concurrent transmissions by geographically separated nodes and the capability of immediate ring access under low and medium load conditions. Traffic regulation by the MAC protocol becomes important for maintaining fair access with tight delay bounds during periods of sustained heavy load.

Protocol efficiency is shown by its capacity utilization compared to the optimum values achieved by OPT. The performance depends on the duration of protocol control cycles. Finer granularity of network control leads to higher traffic regulation since the protocol reacts to 'short-term' unfairness which would be compensated in longer control intervals. On the other hand, the coarser the granularity of network control, the longer a blocked station has to wait for medium access.

CRMA-II offers a scheduling algorithm which results in dynamically changed control cycles in order to minimise the overhead. An enhanced version of this scheme has been used in our investigations [19]. In quota-based protocols larger transmission quota automatically leads to smaller frequency of renewing processes. The values of the used quota parameters have been derived from results of investigations presented in [20]. A window size of 512 slots is chosen for the ATMR protocol, with the Min and Max-quota of the MetaRing set to 8 KBytes.

## 5.1  Saturated Traffic Scenarios

The capacity of slotted rings depends on the actual traffic distribution and increases with higher locality of the traffic flow. Thus, we discuss the capacity loss caused by the MAC protocols versus changing locality of the traffic distribution.

Figure 7 shows the throughput of the protocols in comparison to OPT on a dual counter rotating ring with 10 stations. All protocols are able to provide fairness and high efficiency in bandwidth utilization in all traffic scenarios. The obtained throughput follows the increasing network capacity caused by higher values of p. The MetaRing protocol comes close to the values given by OPT. Its continuously circulating SAT-signal provides an immediate traffic regulation, while the mechanisms of ATMR and CRMA-II have to react on changing load conditions first.

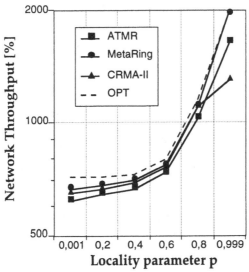

**Figure 7**   Bandwidth utilization vs. locality of the traffic distribution

Compared with OPT performance loss caused by traffic regulation is highest for the uniform traffic distribution, i.e. p = 0,001. OPT does not have to perform any action. All other protocols cannot recognize the global fairness characteristics and have to react within each control cycle. This becomes different for p = 0,999 which also models a symmetric traffic scenario. MetaRing recognizes the fairness characteristics and its performance meets that of OPT. Contrary to that the performance of CRMA-II decreases. Due to the propagation delay the control signal which selects the status information of the network arrives at every station at a different time. Since a station's status is changing rapidly in that particular scenario, the information which reaches the scheduler does not reflect the actual network situation. The scheduler gets the incorrect impression that stations which have been visited by the control signal at the end of its round trip get more bandwidth. Thus, some traffic regulation occurs though if the traffic distribution is symmetric.

Even if the control window of the ATMRing (512 * 48 Byte) is larger than that of MetaRing (8192 Byte) the throughput of ATMR will be smaller. Furthermore, this larger control window leads to higher delays as shown in figure 8. CRMA-II operates more independently of the traffic distribution resulting in some drawbacks for higher locality. Considering the traffic distribution, increasing locality due to higher values of p should reduce End-to-End delays. But only MetaRing behaves as expected. In the other protocols this advantage cannot be preserved due to additional traffic regulation which becomes necessary to compensate the higher asymmetry of the traffic distribution.

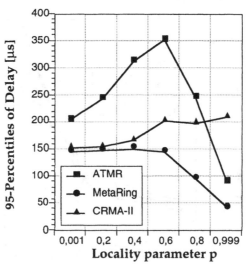

**Figure 8**   Packet delay vs. locality of the traffic distribution

For higher data rates or for geographically extended networks we have to take into consideration the effect of latency due to the finite speed of light. The assumption that the propagation delay is usually smaller than the transmission time is reversed for future networks. Algorithms for congestion control, buffering, and media access control have to be redesigned with particular emphasis to the fact that necessary control information may not be available in due time.

The performance evaluation for increasing ring lengths is restricted to a traffic distribution given by p = 0.5. As shown in figure 9 high efficiency in bandwidth utilization can be preserved for a ring length up to 20 km. For small networks quota based protocols yield better performance. But if the network extends their efficiency decrease. MetaRing behaves particularly bad. This may be explained by the similarities of its traffic control and the token mechanism. The mean rotation time of the SAT-signal is almost constant for short ring lengths and mainly determined by the holding times of the stations. For larger rings the propagation time of the SAT-signal is longer than the time stations need to use their quota. Thus, every station is already blocked when it receives the SAT-signal. During this time gap all slots have to be passed unused. A possible solution is to increase the transmission quota. But higher quota result in a worse timing behaviour as shown in [20]. ATMR also suffers for increased propagation times of the control signals due to higher ring latency. Its performance does not decrease as strong as in MetaRing since we assume twice the quota in the ATMRing.

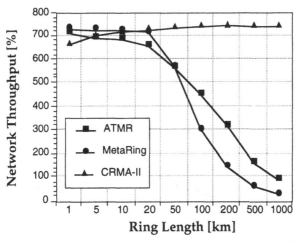

**Figure 9**  Protocol performance vs. ring length, p = 0,5

The reservation-based protocol of CRMA-II is less influenced by increased ring latency. Scheduler processing is detached from data transfer and full network utilization is maintained during traffic regulation.

## 5.2  Non Saturated Traffic Scenarios

The assumption of saturated load conditions in local communication systems is unrealistic. Therefore, we present results which have been obtained in variably loaded scenarios. In figure 10 user throughput provided by ATMR, CRMA-II, and MetaRing is shown for different values of the locality parameter p versus the generated load. It should be noted that the presented results are close to results which have been obtained for values of p up to 0,8. The results show that under low and medium load conditions all protocols are able to handle the offered load. Increasing load saturates the network and no more throughput increase is possible. Then again, all protocols are able to provide global fairness to the competing stations. The points of saturation depend on the locality of the traffic distribution given by parameter p. In each case MetaRing provides the best performance. CRMA-II comes next with the exception for p = 0,999. Under realistic load conditions all protocols succeed in exploiting more than 70 % of the theoretical capacity given by OPT. Compared to the uniform traffic distribution higher asymmetry of the traffic flow for p = 0,8 leads to larger differences of protocol performance and the OPT values. Superiority of MetaRing becomes most clear for p = 0,999 since its continuously circulating control signal is able to renew the transmission quotas in due time which expires very fast due to the extended transmission capacity. Most significant is the performance loss of CRMA-II.

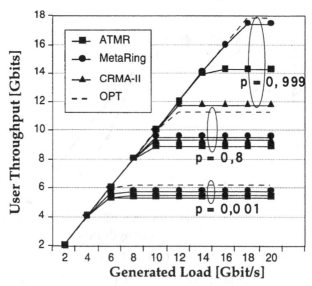

**Figure 10**  User throughput vs. offered load

Figure 11 shows that the delay characteristics of all protocols closely correspond to the obtained throughput. Again MetaRing is best, followed by CRMA-II. Both metrics, the average values and the 99-percentiles, behave quite similarly. The small differences indicate decreased delay jitter; this is in accordance with small values of the coefficients of variation measured in all experiments. The results in figure 11 show that all protocols allow immediate access to the medium up to saturated load where End-to-End delays are mainly determined by waiting times in the transmit buffer. Although delays grow up strongly they remain bounded, and every station is allowed to send during a certain period of time.

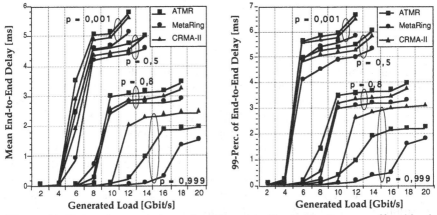

**Figure 11**  Delay characteristics under different traffic distributions vs. offered load

Finally, we discuss the service provided by traffic regulation schemes for different stations in a scenario with p = 0,2 and a load of 8 Gbit/s.

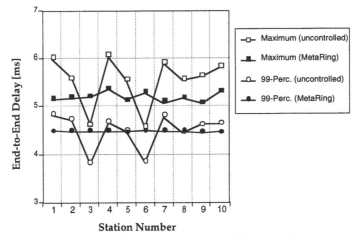

**Figure 12**  End-to-End Delays for different stations

Fair bandwidth sharing is important for network operation even more if some nodes would starve without traffic regulation. But jitter and packet delay become too high if stations get bandwidth 'on block'. Thus, an additional requirement is to achieve acceptable delay characteristics for every station. Figure 12 indicates that MetaRing improves 99-percentiles and maximum values of the delays independently of the traffic distribution. Contrary to that delays are influenced by actual load conditions without any control. It should be noted that the given traffic distribution is moderate because every station gets some bandwidth without the traffic control. In scenarios where some stations are blocked differences would become more obvious.

# 6 CONCLUSIONS

Advancing communication technology and more demanding user applications stimulate the development of new network architectures. A cell based system is favoured in order to take full advantage of growing transmission rates and to provide a broad range of services. Thus, the slotted ring combined with spatial slot reuse is considered as appropriate approach for data transmission and admission control in distributed LANs. By this scheme, the obtainable network throughput may be a multiple of the nominal data rate. In this manuscipt, the starvation problem in such networks has been addressed for different MAC protocols. Major results of the performance evaluation are summarised in the following table:

| | ATMRing | | CRMA-II | | MetaRing | |
|---|---|---|---|---|---|---|
| | Throughput | Delay | Throughput | Delay | Throughput | Delay |
| low load, any traffic distribution | + + | + + | + + | + + | + + | + + |
| moderate load, uniform, low locality | + | o | + | o | + | + |
| moderate load, uniform, high locality | o | o | – | – | + | + |
| moderate load, asymmetric | + | o | + | + | + | + |
| saturated, uniform, low locality | o | o | + | + + | + | + + |
| saturated, uniform, high locality | + | + | – – | – – | + + | + + |
| saturated, asymmetric | + | – – | + + | o | + + | + |
| increased ring length | – | – – | + + | + + | – – | – |
| simplicity of protocol operation | + | | – | | + + | |
| reliability of protocol operation | + | | – | | + | |
| flexibility of bandwidth allocation | – | | + + | | o | |

It is shown that quota-based mechanisms as well as reservation-based protocols with explicit backpressure are suitable to provide fairness. They minimise the tradeoff between high throughput and strong real time requirements, providing high bandwidth for user data and low End-to-End delays. MetaRing was shown to be best under all load conditions. It is able to regulate the traffic flow without a substantial capacity loss and the delay bounds are most efficiently reduced. The drawback of MetaRing is its less efficient operation for larger ring latencies.

In their actual versions the protocols cannot guarantee constant bandwidth connections under high load. Extensions have to be developed to support isochronous services which are provided by ATM technology due to the concept of virtual End-to-End connections. Further work includes protocol optimisation for requirements of specific services as well as dynamic parameter tuning according to changing network states.

Development of LANs is influenced by a significant commercial support in standardisation of ATM LANs and by the vision of an universal ATM based network. Switched-based networks which use trunks for host interconnection might also be suitable for local infrastructures. For a number of economic and practical reasons there may be a coexistence of shared media and cell based switching systems. Thus, substantial efforts should be made to provide interoperability of both architectures.

# REFERENCES

[1]  Kleinrock, L., "The Latency/Bandwidth Tradeoff in Gigabit Networks", IEEE Communications Magazine, April 1992

[2]  van As, H.R., "The Evolution Towards High Performance LANs and MANs", it+ti, vol. 35, no. 4, August 1993

[3]  Kung,H.T.,"Gigabit Local Area Networks: A Systems Perspective", IEEE Communications Magazine, April 1992

[4]  IEEE Network Magazine, "Focus on ATM LANs", February 1993

[5]  Fink, R.L.; Ross, F.E., "Following the Fiber Distributed Data Interface", IEEE Network Magazine, Febrary 1992

[6]  Kasahara, H. et al., "Distributed ATM ring-based switching architecture for MAN and B-ISDN access networks", Proc. of first IFIP Conference on Broadband Communications, Estoril 1992

[7]   Okada, T. et al., "Traffic Control in Asynchronous Transfer Mode", IEEE Communications Magazine, September 1991

[8]   van As, H.R. et al., "CRMA-II: A Gbit/s MAC-Protocol for Ring and Bus Networks with Immediate Access Capability", Proc. of EFOC/LAN, London 1991

[9]   Lemppenau, W.; van As, H.R.; Schindler, H.R., "A 2.4 Gbit/s ATM Implementation of the CRMA-II dual-ring LAN and MAN", Proc. of EFOC&N, The Hague 1993

[10]  Cidon, I.; Ofek, Y., "Metaring - A Full-duplex Ring with Fairness and Spatial Reuse", Proc. of IEEE INFOCOM´90, San Francisco 1990

[11]  Chen, J.; Ahmadi, H.; Ofek, Y., "Performance Study of the MetaRing with Gb/s links", Proc. of IEEE Conference on Local Computer Networks, Minneapolis 1991

[12]  van As, H.R. et al., "Performance of CRMA-II: A Reservation-Based Fair Media Access Protocol for Gbit/s LANs and MANs with Buffer Insertion", Proc. of EFOC/LAN, Paris 1992

[13]  Ito, T. et al.. "Performance Analysis of a High-Speed Ring Network in a Multi-Services Environment", Proc. of 11th International Conference Computer Communication, Genoa 1992

[14]  Bach, C.; Grebe, A., "Comparison and performance evaluation of CRMA-II and ATMR" Proc. of IFIP 5th High Performance Network Conference, Grenoble 1994

[15]  Meuser, T., "Performance Evaluation of the CRMA-II Protocol in Gbit/s LANs", Proc. of EFOC&N, The Hague 1993

[16]  Marsan, M. et al., " On the Performance of Topologies and Access Protocols for High Speed LANs and MANs", Computer Networks and ISDN Systems, vol. 26, no. 6-8, March 1994

[17]  Falconer, R.M.; Adams, J.L.: Orwell, "A Protocol for an Integrated Service Local Network", in "Advances in Local Area Networks", Kümmerle et al. (ed.), IEEE Press, 1987

[18]  Davids, P., "ATLAS - Reference Manual", Department of Computer Science, Aachen University of Technology, Version 5.0, 1991

[19]  Meuser, T., "Enhancement of the CRMA-II Protocol by Distributed Marking", Proc. of EFOC&N, Heidelberg 1994

[20]  Breuer, S.; Meuser, T., "Enhanced Throughput in Slotted Rings Employing Spatial Slot Reuse," Proc. of IEEE INFOCOM ´94, Toronto 1994

# 3

# IMPLEMENTATION AND PERFORMANCE ANALYSIS OF A MEDIUM ACCESS CONTROL PROTOCOL FOR AN ATM NETWORK

## T. Apel*, C. Blondia**, O. Casals***, J. García***, K. Uhde*

*Mikroelektronik-Anwendungszentrum Hamburg,*
*Karnapp 20, 21079 Hamburg, Germany*
*** University of Nijmegen, Computer Science Department,*
*Toernooiveld, 1, Nl-6525 Nijmegen, The Netherlands*
**** Polytechnic University of Catalonia, Computer Architecture Department,*
*c/ Gran Capitan, Modulo D6, E-08071 Barcelona, Spain*

## ABSTRACT

The paper presents the Medium Access Control (MAC) protocol which is being implemented in the demonstrator of the European reseach program RACE project R2024 Broadband Access Facilities (BAF). This MAC protocol is used to concentrate ATM user traffic in a Passive Optical Network with a tree structure. Access to the medium is controlled by means of a request/permit mechanism. The bandwidth allocation algorithm approximates a global FIFO strategy and enforces a spacing of the user traffic. In order to guarantee a limited reaction time, the notion of Request Block is introduced. The MAC chip design is described and the performance of the protocol is evaluated obtaining indications about the transfer delay and the Cell Delay Variation introduced by the protocol and the amount of buffer space needed at the user side.

## 1 INTRODUCTION

Network architectures that allow sharing of the access resources can reduce the costs of the investments needed for the introduction of B-ISDN, particularly for small bussiness and residential customers. Recent progress in the area of optical transmission technology (optical fibers, passive optical splitters and combiners) make a Passive Optical Network (PON) a good candidate for such an access medium. Using this approach, multiple users can share a common optical medium for the interconnection with an ATM Local Exchange.

The European research RACE project BAF (Broadband Access Facilities) is performing theoretical studies and implementing a demonstrator to investigate the problems that arise in the design of such access networks. The passive optical tree has been identified as the most appropriate topology for implementation in the BAF project after comparison with the single star, double star and bus topologies. From the traffic performance point of view, the most important item is the definition of a mechanism controling the access of the different users to the shared medium avoiding collisions.

In this paper we describe this mechanism, called Medium Access Control (MAC) protocol. The MAC protocol we present was chosen among other options ([3]) to be implemented in the demonstrator system. The MAC protocol designed for a PON with a tree structure has centralized control and is defined by:

- The way the central control is informed about the state of the Network Terminations (NT1) (e.g. the number of cells that are waiting in the buffer of the NT1)

- The way the NT1s are informed when a cell can be sent (i.e. permission to access the medium)

- The way the bandwidth is distributed among the NT1s.

The MAC protocol uses a request/permit mechanism. Each NT1 which shares the access network declares its required bandwidth by sending *requests* to the master of the protocol located in the Line Termination (LT) (at the root of the tree). The MAC protocol allocates the available bandwidth to the NT1s according to a *permit distribution algorithm* based on the received information from the requests. The NT1s are informed about this obtained bandwidth by means of *permits*. Such a permit authorizes the NT1 to send a cell. The MAC protocol proposed is cell-based, meaning that an issued permit refers to the transmission of a single cell. The proposed permit distribution algorithm (or bandwidth allocation algorithm) approximates a global FIFO strategy and enforces a minimum distance between consecutive cells of an NT1.

Section 2 presents the architecture considered as the reference for the MAC protocol definition. Section 3 describes the MAC protocol in detail. Section 4 is devoted to the description of the implementation of the MAC chip for the BAF demonstrator system. In section 5 we evaluate the protocol obtaining as performance measures the mean and the variation of the transfer delay, the Cell Delay Variation (CDV) introduced and the distribution of the queue length in the buffer needed at the NT1s. Finally, conclusions are drawn in section 6.

## 2 REFERENCE ARCHITECTURE

The reference configuration is shown in Figure 1. The system uses a Passive Optical Network with splitting factor 32 and maximum of 10km fiberlength. The BAF Network Termination (BAF-NT) contains the Optical Network Unit (ONU) and one or more Network Terminations (NT1), each providing one $T_b$ interface. The Optical Distribution Network (ODN) is realized with a passive optical tree. The BAF Line Termination (BAF-LT) contains the permit distribution algorithm. The following assumptions are done:

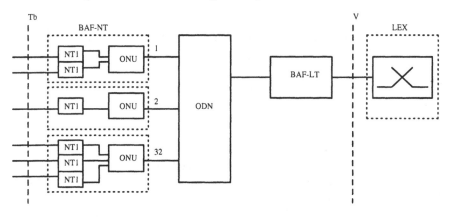

**Figure 1** Reference configuration

- Maximum number of Tb interfaces: 81

- Maximum number of BAF-NTs: 32

- Tb interfaces net bit rate: 149.76 Mbit/s

- V interface net bit rate: 599.04 Mbit/s

- PON internal gross bit rate: 622.08 Mbit/s

- Round-trip propagation delay: $100\mu s$ (assuming a maximum distance of 10km)

# 3 THE MAC PROTOCOL

Since the MAC protocol has centralized control located in the LT, there has to be a means for information exchange between the NT1s and the LT. To achieve this information exchange, a request/permit mechanism is used. The NT1s declare their bandwidth needs by means of *requests*. Based on the information contained in these requests and some additional information concerning the traffic carried by the NT1s, a *bandwidth allocation algorithm* determines the amount of bandwidth each NT1 is allowed to use. The NT1s are informed about this bandwidth by means of *permits to send a cell*. In what follows we discuss these components in detail. First the mechanisms are described and later on, implementation issues will be discussed.

## 3.1 Information Flow between LT and NT1

### Requests
The amount of bandwidth needed by an NT1 can be expressed as the number of cells that are waiting in its buffer. In this protocol, an NT1 has two possible means of informing the LT about this number.

### (i) Requests coupled with upstream cells
Each time an NT1 is allowed to transmit a cell, a tag is added to this cell, containing the number of cells that are waiting in the buffer. Using this type of request, the MAC protocol may react slow on changing traffic situations. In the extreme case, when a cell (or burst of cells) arrives at an empty NT1 buffer, the NT1 has no means to inform the LT about the arrival of this cell, since it has no possibility to transmit a request with an upstream cell.

### (ii) Request Blocks
In order to avoid conditions as described above, a second means to inform the LT about the bandwidth needs is introduced, namely the *Request Blocks* (RB). An RB consists of a series of requests originating from a number of consecutive NT1s. An RB takes the place of a single upstream ATM cell. The way an NT1 is informed that it can put a request in an RB will be treated in the next

paragraph on permits. In order to avoid bandwidth waste, the RBs are only used when no NT1 is allowed to send an ATM cell. Hence, RBs are transmitted using the spared bandwidth between NT1s and LT. The frequency by which RBs are issued is important for the reaction speed of the MAC protocol. A detailed performance study will be made in Section 5.

**Permits**

Upstream information takes two different forms : ordinary ATM cells or Request Blocks. For each type, the LT issues different permits.

### (i) Permits for ATM cells

When the LT decides that an NT1 can transmit a cell, it generates a *permit for an ATM cell*. This permit contains the address of the NT1 that is allowed to send a cell. Such a permit is transmitted downstream in the next time slot, together with an ATM cell or empty cell. Downstream information is broadcasted to all NT1s connected to the LT, hence the NT1 that recognizes its address in the permit, is allowed to send a cell in the next upstream slot (together with a request containing the number of cells left in its buffer).

### (ii) Permits for a Request Block

When no permit for an ATM cell is generated, the LT issues a *permit for a Request Block*. Such a permit contains the address of the first NT1 in a series of $K$ consecutive NT1s that can use the RB. The permit for an RB is sent to the NT1s in the same way as a permit for an ATM cell. When the NT1s recognize a permit for an RB, they compare their address with the address in the permit. If the address of an NT1 is one of the $K$ consecutive NT1 addresses, then this NT1 is allowed to issue a request in the next upstream slot. The comparison of addresses also results in the exact place of the NT1's request in the RB. The $K$ requests are merged together at the optical merger and form an RB (using the same bandwidth as an ordinary ATM cell).

## 3.2    Bandwidth Allocation Algorithm

Based on the information contained in the requests and on information concerning the traffic carried by the NT1s, the LT uses a bandwidth allocation algorithm to decide to which NT1 the next permit will be allocated. This algorithm has to fullfil the following requirements :

- *Fairness* : no NT1 should be favored.

- *Performance* : the traffic characteristics according to which cells enter the NT1 should be kept unchanged as much as possible. In particular, the algorithm should not introduce unacceptable cell delay variation (CDV) on the traffic streams. The cell transfer delay should also be small.

Keeping these requirements in mind, the algorithm used in this MAC protocol has two important characteristics :

- *Spacing function* : Cells originating from the same NT1 are spaced.

- *FIFO strategy* : Cells are transmitted according to an approximate global FIFO strategy.

**Spacing Function**

To perform its function correctly, the LT should have an up-to-date knowledge of the state of each NT1 of the BAF. However, due to the reduced value of the MAC overhead, the LT maintains a distorted image of the state of the NT1s. The NT1s cannot send to the LT information about their state until a permit for a cell or for a Request Block is adressed to them. Hence the LT does not know the exact instant when new cells arrive at a given NT1. Moreover, the LT does not know which particular Virtual Channels (VC) have generated those cells.

If high bit rate VCs are multiplexed in an NT1, it may happen that the same request (coupled with an upstream cell or included in a Request Block) contains information about the arrival of two or more cells of the same connection. If the LT sends directly permits to the NT1s, without introducing any aditional space between them, it could happen that the VC would emit two or more cells consecutively without respecting the maximum peak bit rate of the connection (clumping effect), causing a degradation of the Quality of Service (QoS) seen by other users of the network.

The MAC protocol uses a spacing function for the global traffic emitted by an NT1 in order to avoid these clumps of cells of the same VC. To do that, the LT maintains for each NT1 the number of cells waiting in the NT1 queue for which no permit has been generated yet, the so-called *request-counter (REQ_CNTR)*. When a request from an NT1 arrives, the LT computes from its contents the number of cells that have arrived at that NT1 since the last request (i.e. the number of new arrivals since the last update of the request-counter). This number is then added to the current value of the request-counter. The algorithm then transforms the requests in the request-counter into permits, one at a time, but such that the time between two consecutive permits for a given $NT1 - i$ is at least a certain value $T_{sp}(i)$. In this way, upstream cells transmitted by the $NT1_i$ will be spaced with an interdeparture time of at least $T_{sp}(i)$.

The mechanism described above performs a 'bundle' spacing function, as no individual VCs are spaced, but instead the traffic multiplexed in the NT1. Therefore we should expect that the CDV of a given VC connection at the output of such spacer will be greater than the CDV at the output of a VC-based spacer mechanism (as the one described in [4]). On the other hand, we should expect that the cost of such bundle spacer will be lower than the cost of a VC-based spacer.

The bundle spacer has been introduced to deal with the problems caused by high bit rate connections. Let us consider now the case of a CBR connection of period $T$ with a low bit rate (i.e. with a large value of $T$), multiplexed in the same NT1 with other low bit rate VCs. In this case the MAC protocol cannot introduce clumps of cells of the same VC. Due to the large time interval between the arrival of two consecutive cells of the same connection a VC will not be able to emit two cells consecutively since during time T a number of permits and requests from this NT1 are issued. Hence no spacer is required. On the other hand, if we try to perform the spacing function on these low bit rate connections based on the peak bit rate of the superposition ($T_{tot}$), i.e. fixing $T_{sp} = T_{tot}$, we can have large values of CDV and transfer times. To explain this, consider the case in which we have a low bit rate CBR connection with period $T$ multiplexed in the same NT1 with another connection with a different but similar low bit rate. If we fix $T_{sp} = T_{tot}$, $T_{sp}$ has now a large value (close to $T/2$). If one cell of the second connection arrives between two consecutive arrivals of the tagged connection, the distance between two consecutive departures of cells of our tagged connection is $2T_{sp}$, which is close to $T$. However if we do not have any arrival, or if we have two arrivals of the background source, the interdeparture time will be $T_{sp}$ or $3T_{sp}$. Then in this example we have that the bundle spacer can introduce values of CDV or waiting times of the order of $T/2$, which can be of several millisecs. To avoid this situation we should fix $T_{sp} < T_{tot}$. Hence, we have that *the optimal value $T_{sp}$ does not depend only on the peak bit rate of the total traffic, but also on the characteristics of the individual connections.*

It is desirable to have a simple rule to fix the value of $T_{sp}$, while obtaining low values of CDV and transfer times. The rule which has been chosen and which only needs to know the sum of the peak bit rates per NT1 is the following:

$$T_{sp} = min(T_{tot}, T_{max}). \qquad (3.1)$$

The value $T_{max}$ has to be chosen properly so that it does not cause cell clumping in the high peak bit rate connections and allows low values of CDV and transfer times in the case of low bit rate connections. This rule is evaluated in section 5.

**Global FIFO Strategy**

It is possible that at the same instant, two or more permits for different NT1s are generated, which will compete for the same downstream time slot. In this case, only one permit can be transmitted and the others have to queue. The LT is provided with a permit FIFO buffer, in which permits queue according to a FCFS strategy for being transmitted to the respective NT1. Such a strategy ensures fairness and minimal delay variance. Hence, using this mechanism, cells wait in the NT1 buffers, which are controlled by the LT (distributed buffering with central control).

## 3.3 Robustness of the MAC Protocol

In order to operate correctly, the NT1s must declare their bandwidth requirement (using requests coupled with upstream cells or using Request Blocks), through the number of newly arrived cells since the last request. However, the loss of a request may lead to the situation that cells remain in the NT1 buffer forever. In order to avoid this situation, the request contains the queue length of the NT1. The MAC controller must be able to compute the number of new arrivals from this queue length. In order to do so, *mirror counters* are introduced [2]. The *mirror counter* $M_i$ maintains the buffer queue of $NT1_i$ after the last request has been transmitted. When a permit is lost, the system will be able to recover from this loss as the LT will be informed about the presence of the cell for which the permit was lost by means of one of the next requests or request blocks, since a request contains the NT1 queue length. Using the mirror counter scheme, loss or corruption of permits or requests can lead to an increase in CDV but not to a failure of the MAC protocol leaving cells in the NT1 buffer forever.

# 4 IMPLEMENTATION OF THE MAC PROTOCOL

Now we show how the mechanisms of the MAC protocol, as defined above, are implemented.

## 4.1 Information Structure

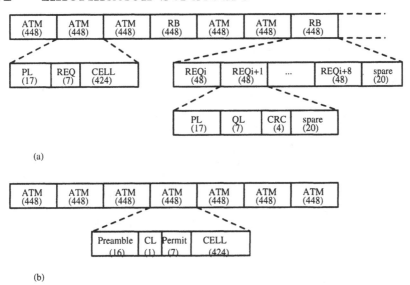

(a)

(b)

**Figure 2** Information Structure: (a) Upstream, (b) Downstream

**Upstream Information Structure**
Figure 2a shows the structure of the upstream information flow. A slot containing an ATM cell has a length of 448 bits, consisting of a physical layer preamble (17 bits, for physical layer overhead), the request itself (7 bits, for the NT1 buffer length) and the ATM cell (424 bits).

A slot containing an RB has the same length and consists of 9 request fields. A request field contains again the physical layer preamble (17 bits, needed since each request originates from another NT1), the request itself (7 bits, for the NT1 buffer length), an error detection code (4 bits, CRC able to detect and correct 1 bit errors) and 20 bits spared. In the demonstrator an RB contains 9 requests.

**Downstream Information Structure**

Figure 2b shows the structure of the downstream information flow. A slot has a length of 448 bits and consists of a physical layer preamble of 16 bits, a (CL) bit indicating whether the permit that is added refers to a permit for an ATM cell (CL=0) or a permit for a request block (CL=1), the permit (7 bits being the address of the NT1 which is allowed to send a cell (if CL=0) or being the address of the first NT1 in a series of 9 consecutive NT1s, which are allowed to send a request in order to form a request block) and finally the downstream ATM cell (424 bits).

## 4.2   The MAC Chip

The bandwidth allocation algorithm will be implemented in a dedicated hardware. In order to achieve a reliable and fast design flow, a CMOS technology with a 'standard cell' approach has been chosen to realize the MAC chip. In this case, predesigned 'cell' provided by the manufacturer as library of different CMOS gates are used to design the circuit. This yields high flexibility but avoids the efforts in design and time of a 'full custom solution' where the smallest design unit would be e.g. a simple transistor. To cope with a high complexity of the ASIC a 0.8 $\mu$m CMOS process has been chosen. As the MAC chip will be used in a complete PON system for shared access, it contains the Permit Distribution Algorithm (PDA) and two additional functions which will be treated later on. Figure 3 shows a block diagram of the MAC ASIC. The core of the Global FIFO PDA as given above, i.e. the generation of cell permits, can roughly be divided into three parts:

- The incoming queue length as well as address information are processed and the number of new cell arrivals is calculated by making use of a mirror counter algorithm. This is implemented in the 'data-in processor' block and the 'mirror counter RAM'.

- For each NT1, permit flags are generated based on a countdown process which is controlled by the request counter; the request counter value is updated when there are 'new arrivals' data from the 'data-in processor' block, or when a permit flag is generated. This functionality is realized in the PDA block.

- The third part deals with the processing of the permit flags and the generation of NT1s addresses for the cell permits. This is implemented in the 'group permit generator', the 'group permit FIFO' and the 'group permit processor'. The cell permits are stored in the Global permit FIFO.

Additionally, the MAC ASIC will contain a 'permit selector'. It takes normal cell permits out of the Global permit FIFO or generates RB permits as described above, plus periodical operation and maintenance (OAM) permits as well as special permits for ranging purposes on demand. As the system supports Fiber To The Home (FTTH) and Fiber To The Curb (FTTC), the OAM and ranging permits will be transmitted only once per network termination/unit (NT) independently how many subscribers it supports. Therefore, these permits have a special address which differs from the normal NT1 interface address. To distinguish it from the MAC protocol endpoints, the abbreviation 'NT' will be used in the following.

Each generated and transmitted permit is also stored in a FIFO, the upstream indication FIFO. It serves as a reference for the complete LT and provides the MAC chip with the NT1 address which the incoming queue length information belongs to. Via the $\mu P$ interface control information is passed from the LT to the MAC chip.

### The Data-in Processor
After a request or a request block has been received, the data-in processor calculates the number of newly arrived ATM cells by the use of a mirror counter algorithm. An equivalent of the queue length information an NT1 has sent in its last request is loaded in a RAM, in the so-called mirror counter RAM. To increase the overall access time, a dual-port RAM is taken using one port to read the mirror counter value into the RAM and the other port to write the up-to-date mirror counter value into the RAM. The mirror counter value is equal to the last queue length information decremented by one in case the last request was carried by a normal cell; in case of an RB the unmodified queue length information is kept. Thus, the actual buffer queue of a NT1 after having transmitted the last request is tracked. The difference between the incomming data $QL_i$ and the value of the mirror counter $M_i$ determines the number of new arrivals at $NT1_i$. This value is forwarded to the request counter for $NT1_i$. Assuming a corrupted queue length information has been received, the number of new arrivals is set to zero. Additionally, the mirror counter value is decremented if the request was transmitted by a cell.

In case of an RB, the data-in processor also calculates the addresses of the NT1s from the address information given in the permit.

### The PDA Block
In order to implement the permit distribution algorithm (PDA), three registers per subscriber/ NT1 are necessary. The first register is needed to implement the request counter of $NT1_i$. It holds the number of cells that are waiting in

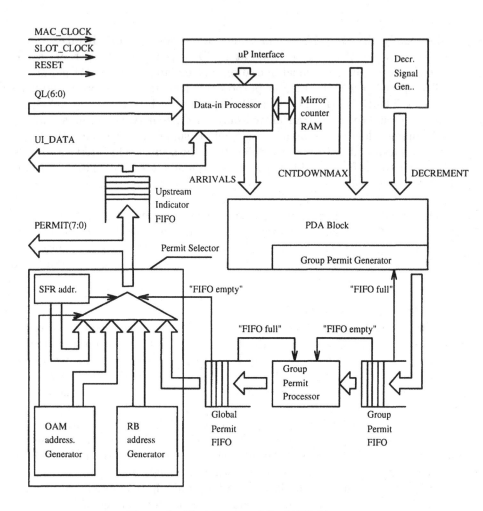

**Figure 3**   Blockdiagram of the MAC chip

the buffer of $NT1_i$ and for which no permit has been generated yet. Its value is denoted by $REQ\_CNTR(i)$. It is increased by the number of new cell arrivals after a request of the corresponding NT1 has been received and it is decreased by 1, whenever a permit for $NT1_i$ is generated. The spacing function is realized by the use of a second register denoted by $CNTDN\_CNTR(i)$. Together with a decrementer it forms the count-down counter. If the spacing function requires $T_{sp}(i)$ slots between two consecutive cells from $NT1_i$, the register is loaded with the value $T_{sp}(i)$, each time a permit for $NT1_i$ is generated. Then, it is decremented once per time slot until its value becomes less or equal zero. If at this moment $REQ\_CNTR(i)$ has a value different from zero, a permit flag for this NT1 is set. Additionally, $CNTDN\_CNTR(i)$ is reset to $T_{sp}(i)$ and the count-down process starts again. If on the other hand the request counter is zero at that moment, then the count-down process goes sleeping. It will be awaked when a request from $NT1_i$ with a value greater than zero arrives. In this case a permit is generated immediately. As each $NT1_i$ is specified by a value $T_{sp}(i)$, the third register is needed to store the value $T_{sp}(i)$. Its value will be updated in case the bandwidth allocation of $NT1_i$ changes.

The spacing period $T_{sp}(i)$ the count-down counter of $NT1_i$ is loaded with is computed, using 3.1, from the peak bit rate of all connections $NT1_i$ carries. If the count-down counter is decremented once per time slot then a value $T_{sp}(i) = 1$ is given to an NT1 which uses about the overall bandwidth of the system; $T_{sp}(i) = 2$ denotes an NT1 using half of the available bandwidth, etc..., while a large value of $T_{sp}(i)$ indicates that the bandwidth allocation of $NT1_i$ is small. Thus, calculating $T_{sp}(i)$ the truncation error increases with increasing bandwidth allocation of $NT1_i$ which may result in a waste of bandwidth. One possibility to increase the accuracy of coding the spacing periode is to decrement the count-down counter more than once per slot while the original value $T_{sp}(i)$ has to be multiplied with the corresponding factor. This reduces the granularity for all NT1s independently from their peakrate but increases the size of the registers i.e. the area on silicon as well as the overall power consumption.

Looking for a compromise we have chosen a coding scheme in which only the count-down counters of NT1s with a high peak rate are decremented more often; they are decremented four times per time slot. The count-down counters of NT1s with medium or small bandwidth allocation are decremented once per time slot or even less depending on their overall traffic. To distinguish between $NT1_i$ with different decrement cycles, two additional bits are used. As the range of $T_{sp}(i)$ to be decoded in one register becomes smaller, it is possible to decrease the size of the registers, to keep the power consumption within an acceptable level and to increase the accuracy of the coded spacing period in case the overall bandwidth allocation of an NT1 is high. The signals needed for the different decrement cycles are generated in the decrement signal generator.

**The Group Permit Block**
Permit flags may be generated four times per time slot. Additionally, several permit flags may be set at the same instant. Therefore, the permit flags of all NT1s have to be read and reset four times within one time slot. As it is not possible to generate permits for, in a worst case, all NT1s in one time slot a so called 'group permit solution' is implemented. A group permit is an

intermediate format in which a number of permits can be encoded. Assuming e.g. an overall bit rate of 622.08 Mbit/s and a MAC clock cycle of 38.88 MHz which means there are 28 MAC clock cycles within one time slot, the number of all possible NT1s is divided into 7 permit groups. The address of such a permit group is coded by three extra bits. In one MAC clock cycle, the permit flags of one permit group are then read. They are stored as one entry in a group permit FIFO together with the permit group address. The permit groups are cyclically served in a fixed round robin scheme. In this way, the permit flags of all NTs are read four times per time slot. In a group permit processor, the permit flags will be taken out of the group permit FIFO and processed one after the other. From the coded permit group address and the position of a flag within one row, the address of the corresponding NT1 can be derived. This is needed to generate a cell permit which is then stored in the Global permit FIFO.

**The Permit Selector and The Upstream Indication FIFO**
Apart from the cell and RB permits, there will be two additional types of permits which are due to system requirements. In order to maintain an exchange of OAM information between LT and NTs, permits for OAM cells will be issued. Having priority over normal cell permits these OAM permits will periodically be inserted in the flow of cell permits out of the Global permit FIFO. The permit issuer will cyclically serve each NT with an OAM permit. The cycle duration is set by the LT and stored within the MAC chip in an internal register. An NT connected to a broadband access via a PON with different fiber lengths, implying a round trip delay which differs from one NT to another, has to be ranged during its installation. Then, the different delay times are compensated by an appropriate delay in the start of transmission of each NT. After a rough adjustment, special permits for a static fine ranging (SFR) process will be issued during the installation of an NT. On the receipt of such a permit, the NT only sends the physical layer preamble. From its position within one time slot, the LT derives control information for the fine adjustment of the NT. The issue of a fine ranging permit is initiated and controlled by the LT. Ranging and RB permits are only inserted when the Global permit FIFO is empty, with the ranging permits having priority.

The permit selector generates RB, OAM and SFR permits, if needed, and merges them with the flow of normal cell permits out of the Global permit FIFO in the given priority. As four types of permits have to be distinguished an appropriate permit coding has been chosen. Normal cell permits and OAM permits are marked by the same CL bit as well as RB and SFR permits by using the upper addressing space for the OAM and the RB permits. Each permit that is generated and inserted in the downstream data flow is additionally stored within the MAC chip in the upstream indication FIFO. By the use of a counter, the system roundtrip delay is emulated in terms of time slots. After one roundtrip delay, the permit is taken out of the upstream indication FIFO. It is used to determine the cell type which will be received in the next time slot and to associate incoming queue length information with the corresponding NT1 address. As reference this information is also provided to the complete LT.

# 5   PERFORMANCE EVALUATION OF THE MAC PROTOCOL

In this section the performance of the MAC protocol is evaluated. Only CBR sources are considered as we assume that in the first deployment phase of the ATM network, peak bit rate reservation will be used as a Connection Admission Control (CAC) algorithm. The following performance measures have been studied:

- the mean transfer delay

- the variation of the transfer delay, evaluated by a quantile of the statistical distribution

- the statistical distribution of the queue length in the NT1 buffer

- the statistical distribution of the Cell Delay Variation (CDV) measured as the difference between a quantile of the maximum and the minimum interdeparture times of two consecutive cells of the same connection.

## 5.1   Evaluation of the Reaction Time

The Request Blocks have been introduced to guarantee fast reaction of the protocol on changing traffic situations. By issuing them during idle periods of the global FIFO queue, no bandwidth is wasted, but an NT1 may have to wait for an idle period during which a permit for an RB, containing the address of that NT1, is generated. This waiting time is called the reaction time. A performance measure of interest for this reaction time is the buffer capacity needed at the NT1 to cope with the reaction time. The time before the first usefull RB permit arrives consists of a number of busy and idle periods of the

permit FIFO queue. We tag an NT1 which generates traffic according to an
on/off source model. Assume the background traffic to be CBR traffic having
the same bit rate and random phases. In [1], a busy period analysis of the
$N \times D/D/1$ queue is used to derive the $10^{-9}$-quantile of the distribution of the
number of cells of the tagged source that arrive in the NT1 buffer before the
first RB arrives. Figure 4 shows this quantile for various bit rates for the tagged
NT1 (34 Mbit/s, 20 Mbit/s, 10 Mbit/s and 2 Mbit/s) under the assumption
of homogenous background traffic ranging from 64 Kbit/s to 20 Mbit/s. The
number of background traffic sources is such that the total load is 0.8. We see
that longer queues are required when the tagged source has a high bit rate and
the background traffic has a low bit rate.

We must point out that the assumption of the CBR background traffic results in
a bounded reaction time (the busy period of the $N \times D/D/1$ queue is bounded).
The case of VBR traffic, currently under study, may lead to longer reaction
times. This problem can be solved by inserting additional permits for a RB in
order to decide the maximum distance between consecutive RBs.

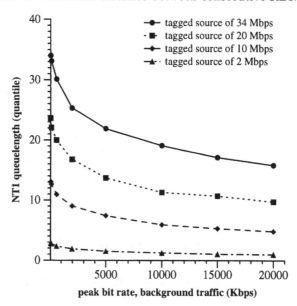

**Figure 5**   Queue length distribution at NT1

## 5.2   Evaluation of the Spacing Function

As any multiplexer, the BAF introduces a certain degree of distortion in the traffic, which is known as Cell Delay Variation (CDV). The BAF is provided with a bundler spacer which tries to avoid the problems caused by an excessive degree of CDV, that could lead to the existence of clumps of cells emitted by a given VC ('clumping effect') which could degrade the QoS seen by other users of the network.

The degree of CDV introduced by the access system (BAF) depends on the value at which the bundler spacer is set $(T_{sp})$ and on the traffic conditions. Using 3.1 to decide the value $T_{sp}$ we have that the critical cases appear when $T_{tot}$ is slightly bigger than $T_{max}$. As we do not distinguish between the case in which the NT1 just serves one connection or a number of low bit rate connections with identical value of total peak bit rate, the same value $T_{sp}$ is used in both cases. If $T_{max}$ has a low value, we can have, in the case of a single connection with a period slightly bigger than $T_{max}$, that excessive cell clumping is introduced by the spacer. On the other hand, a large value of $T_{max}$ would lead to large values of CDV for the case of low bit rate connections.

Let us study a case in which we use $T_{max} = 42\mu secs$. The BAF has 80 active NT1s each one loaded with 6 Mbit/s traffic (that means that the total load of the system is around 0.8). Under this condition we have that $T_{tot} = 60\mu secs$, i.e. we are in a critical case. The 6 Mbit/s traffic in each NT1 can be the result of having a single CBR connection with that peak bit rate value or having a number of lower peak bit rate CBR connections multiplexed in the same NT1. We will study the CDV suffered by a tagged connection multiplexed with other connections.

The maximum degree of variability is achieved when the tagged source is multiplexed with a large number of very low bit rate connections. If the peak bit rates of the sources of the background traffic are much lower than the peak bit rate of the tagged source, we can approximate this background traffic by a Poisson process. In Figure 5 we show values of the $1 - 10^{-3}$ quantiles of the maximum and minimum interdeparture times of the tagged connection at the output of the BAF. The CDV can be measured as the difference between these two values which have been obtained by simulation.

The CDV introduced by the BAF is bigger for low bit rate connections than for high bit rate connections. For example in the case of a 64 Kbit/s connection, the CDV is 522.9 $\mu$secs, for a 128 Kbit/s connection it is 547.4 $\mu$secs while for a 2 Mbit/s connection it is 205 $\mu$ secs and for a (single) 6 Mbit/s connection 70 $\mu$secs. We should point out that for low bit rate connections modelling the background traffic as a Poisson traffic should lead to an overestimation of the CDV introduced by the BAF. For example, in the case of having as background traffic a superposition of 64 Kbit/s sources and a tagged source of 64 Kbit/s we obtain a CDV around 120 $\mu$secs.

On the other hand, the fact of setting $T_{sp}$ to $T_{max}$ can reduce the interdeparture times of the high bit rate connections, so that we can have values of peak bit rate at the output of the spacer bigger than the declared ones. However the probability of having these fluctuations is small, so the QoS seen by other users of the ATM network should not be damaged.

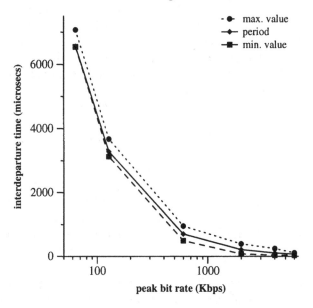

**Figure 6**   Interdeparture times for a CBR source multiplexed with Poisson traffic

## 5.3    Evaluation of the Transfer Delay

One important parameter to assess the performance of the system the distribution of the transfer delay, i.e. the time it takes for a cell that arrived at a given NT1 to be transported to the LT. This performance parameter has been obtained by simulation.

The transfer delay has different components. One component is constant (round trip delay) and the others are variable: the time until a request can be sent, the waiting time in the bundle-spacer and the time in the Global permit FIFO. Consequently the values of transfer delay will depend on the load conditions of the BAF. For example we should expect that when the BAF is heavily loaded, the times until the request can be sent as well as the waiting time in the Global permit FIFO will be higher than for low load conditions. The value $T_{sp}$ will have an important impact on the waiting time in the bundle spacer.

We will study now the transfer delay under the same conditions as the one described in section 5.2. The BAF is working in a critical condition (the load is high and $T_{tot}$ is close to $T_{max}$).

In figure 6 we show the mean value and the $1 - 10^{-3}$ quantile of the transfer delay for a tagged source of different bitrates multiplexed with Poisson traffic. We see that the mean value of the delay is almost constant for different values of the bit rate of the tagged source, and close to 200 $\mu$ secs (150 $\mu$ secs are due to the round trip delay). The quantile of the transfer times is highly dependent on the bit rate of the tagged source, mainly due to the waiting time in the bundler spacer. However the values are bounded by 500 $\mu$ secs in all cases. Again we should point out that modelling the background traffic as a Poisson process can be very pessimistic for low bit rate sources.

# 6   CONCLUSIONS

We have described a MAC protocol for a broadband access facility using a passive optical network with a tree structure. The protocol uses a request/permit mechanism to control the access to the shared medium. The available bandwidth is allocated by means of a strategy which approximates a global FIFO queue and such that a minimum distance between cells is enforced. In order to guarantee a limited reaction time on changing traffic situations, Request Blocks have been introduced. They allow quick NT1 queue status information transfer to the LT. They are scheduled during idle periods of the global FIFO queue. The MAC chip design, as it will be implemented in the BAF system demonstrator has been described. The advantages (low costs) and drawbacks (increase of CDV) of using a bundle-spacer have been analysed showing that an acceptable CDV can be achieved by using a simple rule to determine the spacing value ($T_{sp}$) of the bundle spacer.

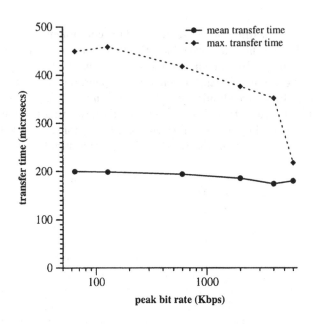

**Figure 7**   Mean and $1 - 10^{-3}$ quantile of the transfer delay

# REFERENCES

[1] O. Casals, J. Garcia, C. Blondia, "A Medium Access Control Protocol for
an ATM Access Network", IFIP Transactions C-21, High Speed Networks
and their Performance, Proceedings of the 5th International Conference
on Data Communication Systems and their Performance, Raleigh, North
Carolina (U.S.A.), October 1993, pp. 289-308.

[2] J. Charzinski, T. Theimer, "A Modified Global FIFO PDA Using Queue
Length Information", Internal document BAF-SWP2-UST-92007-CD-
CC/a, September 1992.

[3] T. Toniatti, L. Verri, O. Casals, J.García, C. Blondia, J. Angelopoulos, I.
Venieris, "Performance of Shared Medium Access Protocols for ATM Traffic Concentration", European Transaction on Telecomunication (ETT),
Special Issue on "Teletraffic research for broadband-ISDN in the RACE
program", vol. 5, No.2, March-April 94, pp. 219-226.

[4] E. Wallmeier, T. Worster, "A Cell Spacing and Policing Device for Multiple
Virtual Connections on one ATM Pipe", Proc. RACE R1022 Workshop
on ATM Network Planning and Evolution, London, 1991.

# FAST RESOURCE MANAGEMENT IN ATM NETWORKS

Pierre E. Boyer

*France-Telecom CNET*

## ABSTRACT

Network resource management is a constant problem, even in high-speed networks. In permanent connections, set-up is done once, which implies allocating some network resources regardless of the real-time traffic profile. On the opposite, switched connections are made only when necessary and break when no longer needed. Therefore, switched connections can be viewed as an approach to network resource management of sporadic traffic profiles. However, switching connections implies the design and deployment of large signalling systems. In addition, the set-up phase duration is not expected to range below a few hundreds of ms which can be far too large for small bursts of activity.

The user of an ATM connection - whether it is permanent or switched - could be alternatively allowed to purchase a temporary increase of network resource. In order to adapt to the real - time traffic profile, the allocation of network resources should be performed within the smallest amount of time - even reducing to the electrical propagation delay. This is the Fast Resource Management (FRM) concept. This paper illustrates some of its early applications which can be viewed as resource switching over permanent connections. But FRM goes far beyond as it applies to switched connections as well and fits the self-similar nature of broadband traffics.

## 1 INTRODUCTION

An ATM-based network primarily offers connection-oriented services : any communication between end users involves a connection establishment phase, a data transfer phase and a connection termination phase. At the end of the establishment phase, the connection is offered a physical path and the allocation of some transmission resources through the network. At the end of the termination phase, the connection path is deleted while network resource allocation is cancelled.

ATM connections come in two types : first, Virtual Path Connections (VPC) identified by the Virtual Path Identifier (VPI) field of the ATM cell header ; then,

51

Virtual Circuit Connections (VCC) identified by the field resulting from the concatenation of the VPI field and the Virtual Circuit Identifier (VCI) field of the cell header.

VP and VC connections can be either permanent or switched. A permanent connection can be obtained upon user subscription. Currently, this is a matter of days ; in the B-ISDN, physical path selection and resource allocation will be performed by the operator via an X-user interface and the Telecommunication Management Network (TMN) ; response time is expected to reduce to minutes. During the connection's life, a re-negotiation of network resource allocation may be performed - with a response time in the range of minutes - to adapt to the real-time traffic profile.

A switched connection is set-up and terminated by means of signalling procedures activated by the user and completed within a very few seconds to be acceptable.

Compared to a permanent service, it takes time to come to a switched service which is far more complex to implement. Indeed, signalling systems are subject to standardization. ITU Study Group 11 is currently working on B-ISDN signaling procedures at the UNI (Q.2931, Q.2962, Q.2963) and at the NNI (Q2761 to 2764). Q.2931 will be the first to be released, by September 1994 ; it addresses the set-up of a single connection declaring a peak cell rate. Q.2963 addresses in-call bandwidth re-negotiation primitives and should be released by mid-1995. Then, many industry efforts are required both in network and customer equipments.

■ Therefore, it can be understood that permanent ATM connections be available first in public carrier services.

At this point, it has to be recognized that ATM service providers have to purchase leased lines from STM (PDH, SDH) infrastructure operators to support ATM connections. To protect his revenues - i.e. avoid bandwidth re-selling - this infrastructure operator cannot price leased lines dedicated to ATM services much lower than the others.

The ATM service provider may find safe and simple to allocate bandwidth to permanent ATM connections according to their declared peak cell rate. But in this case, their pricing would have to reflect the overhead of ATM (cell header) as it cannot be compensated by any statistical gain. As a result, they would be more expensive than leased lines although being in essence functionaly equivalent. Indeed, they would just provide a finer rate granularity and a more flexible quality of transfer than leased lines.

■ Permanent ATM connections which would be allocated their peak cell rate are likely not sufficiently attractive to motivate user subscription :

either he produces smooth traffic - at the expense of some smart multiplexing in its own equipment - and should better subscribe directly for an STM leased line or else he produces bursty traffic and has to pay for the peak cell rate.

However, in principle, ATM services can be based upon some statistical sharing of the network resources. By overbooking the underlying STM leased lines, the ATM service provider can introduce usage as part of service pricing. Compared to flat-rate pricing, users can achieve large savings depending upon real-time traffic profiles. In the USA for example, SPRINT which is the first public carrier to offer long distance ATM connections proposes fixed and usage-based pricing options. Estimating 20% utilization - LAN traffic on a T1 link is usually assessed at peak for only 15% to 20% of the time - SPRINT announces that usage-based pricing should save about 1/3 compared to flat-rate pricing.

- Quality-of -Service (QoS) is the essential issue an ATM service provider is faced to when performing statistical multiplexing :

does he commit himself to achieve any QoS? If so, does he provide for several QoS classes? In this case, how does he achieve their co-existence in the STM infrastructure? How does he plan to maintain QoS committments as ATM user population grows?

The first ATM service providers are unable to commit themselves to QoS requirements. For example, cell loss rate is not guaranteed by SPRINT on any ATM connection. This is not very surprising if one considers the poor available knowledge of broadband traffics and the restricted control capabilities of existing ATM equipments.

Indeed, statistical capabilities to add into the ATM network elements are complex to finalize and are not yet available : they should introduce only a minimum over-head, be acceptable to the ATM users, be low-cost and simple to implement and accept to co-exist. Several proposals are under study and this paper will actually promote one of them.

- Signalling can be viewed as one way to achieve some overbooking of the underlying STM leased lines:

switched ATM connections can be concentrated at the expense of some call blocking. Even if they are allocated their peak cell rate, switched connections are at least set-up only when necessary and terminated when no longer needed. This solution is unanimously favoured by network operators as it is classical, safe and conservative; much efforts are made to build the signalling systems standards.

- However, signalling is not the generic way to achieve a statistical sharing of network resources.

Indeed, the connection set-up duration spreads over seconds which considerably slow down the user access to the network - possibly not compatible with user requested QoS. Besides, a good statistical gain is only achieved if individual bandwidth requirements are low compared to the leased line capacity and if connection durations are sufficiently large - say, at least ten times the set-up duration. Fast Circuit Switching is the limiting resource allocation scheme when set-up is speed-up at its best [1].

■ This is where Fast Resource Management [2] appears:

is it necessary to lose time setting up a new physical path between the two same given endpoints each time before starting any new transmission? Is it necessary to develop costly complex standardized signalling software procedures to indicate just a change of traffic activity on an ATM connection?

A dynamic resource allocation performed within a round-trip electrical propagation delay is sufficient in many situations to adapt the ATM connection to its real-time traffic profile.

## 2  WHAT IS FAST RESOURCE MANAGEMENT IN ATM NETWORKS?

Fast Resource Management is an ATM layer capability to allocate on user request transmission resources to a connection within the shortest amount of time - the target delay being due to the electrical propagation along the round-trip. Resource requests are conveyed by FRM cells identified by a particular payload type code [3].

FRM should not be mistaken with the so-called Burst Switching principle as proposed by S. R. AMSTUTZ in 1983 [4]. Burst Switching is a transfer mode according which digitized voice and data can be conveyed in an integrated way by variable-length packets.

To my knowledge, this idea has been first published by J. HOI ; he recognized that multimedia workstation traffics presented an embedded bursty structure and proposed to use "burst pilot packets" to convey bandwidth requirements to help multiplexing these bursts [5]. In the same issue, OHNISHI and al. proposed a "short-hold mode ATM service" according which the network allocates and de-allocates resources according to user requests conveyed by Q.931 signalling messages [6]. At the same time, we did some hardware investigations about "on the fly" bandwidth reservation in a switching element [7].

We proposed in-call bandwidth negotiation under the name of Fast Reservation Protocol [8] ; DOSHI and al. proposed alternatively in-call bandwidth negotiation of buffer space [9] as well as in-call bandwidth negotiation ; he suggested that network nodes may offer smaller than requested bandwidth - which has something to do with what we call now an ABR service. J. TURNER resumed the idea of fast buffer reservation [10].Then, we specified two tentative protocols namely, FRP/DT and FRP/IT [11].

Performance comparison studies with others statistical capabilities have been also carried by several authors [12,13,14,15,16] which all conclude that there is definitely a place for FRM procedures among broadband network traffic controls.

However, as quoted by R. JAIN in a famous paper [17], traffic control in high-speed networks is a quasi-religious topic - mainly due to our ignorance of ATM connection real-time traffic profiles and application QoS requirements - and FRM has spent some hard times in the telecommunication research community. The last ATM Forum UNI specification does not refer to it [18] and the US delegation in the ITU recently contributed against FRPs [19]. Recently, however, the concept of an Explicit Rate Indication has emerged in the ATM Forum as a basis for traffic control, proposed by XEROX, MIT [20] and British Telecom [21]. This indication is conveyed by FRM cells. Besides, NTT considers that bandwidth on-demand is part of the introduction strategy of broadband services [22].

## 2.1 An Overview of FRP/DT

Fast Reservation Protocol with Delayed Transmission (FRP/DT) [11] is a particular application of FRM addressing in-call negotiation of a peak cell rate . An ATM connection controlled by FRP/DT accepts to wait until reception of a network agreement prior to actually switch to another peak cell rate value.

A network overview is presented (see figure 1). An FRP Unit located on the network side of the UNI checks whether bandwidth requests (REQ and REL cells) conforms with declared traffic parameters. An FRP capability is added to the ATM public network - as already said, it can be either integrated in the VCX or plugged in.REQ cells are discarded by the network element in which the allocation could not be achieved. Otherwise, a timer is set and bandwidth is reserved. If the requested peak cell rate can be allocated in every network element, the egress node sends an ACK cell back to the FRP Unit. Upon reception, the FRP Unit both forwards it to the user as the network agreement and sends a VAL cell to update the value of the peak cell rate in the policing function - UPC(PCR) - and unset timers in the network elements.

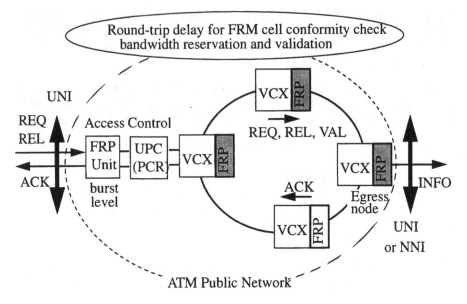

**Figure 1**   Network overview of FRP/DT

## 2.2  FRP/IT

Fast Reservation Protocol with Immediate Transmission (FRP/IT) [15] is a particular case of Fast Resource Management addressing peak cell rate allocation. It is not actually a protocol in the ATM network layer in the sense that there is no control signals exchanged between user and network.

FRP/IT delineates each user data block (referred to as the CS-PDU) by means of two FRP cells, one before the first data cell and one after the last data cell (see figure 2).

The first FRP cell carries the peak cell rate value according which successive cells of the CS-PDU are initially spaced. Transmission is initiated in a network element only if the requested peak cell rate can be reserved on the output port. Otherwise, every cell of that particular CS-PDU is discarded upon arrival in this element. The bandwidth is released upon reception of the last FRP cell.

Therefore, cell loss is concentrated on the last incoming CS-PDU rather than spread over all different CS-PDUs competing for transmission. This is very beneficial to the goodput of the connection in particular when the CS-PDU has to be re-transmitted as soon as one cell is missing upon reception - see AAL5. (see figure 3) below shows a network overview of the FRP/IT.

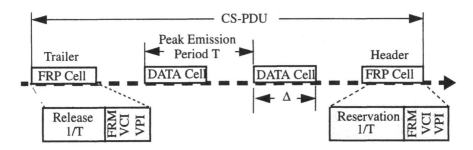

**Figure 2** A CS-PDU delineated according to FRP/IT and transferred along an ATM multiplex.

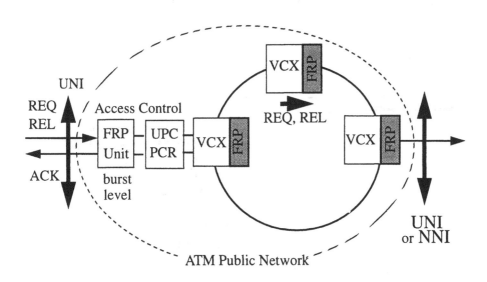

**Figure 3** A network overview of FRP/IT

## 3   A GRACEFUL NETWORK EVOLUTION TO STATISTICAL CAPABILITIES

The first ATM service providers are currently deploying their networks and ATM-native user-network interfaces will be available very soon.

The first ATM network equipments have very poor control capabilities to help performing statistical sharing of transmission resources. They have small buffering capabilities poorly adapted to open-loop rate-based multiplexing or backpressure controls - such as explicit congestion notification.

In the perspective of an ATM operator willing to upgrade his network to statistical capabilities, FRP/DT has a major advantage over backpressure controls : it can be superimposed to the existing ATM network without requesting any logical change nor buffer addition **into** the existing elements already in operation.

## 3.1  Upgrading  a switching  element

For that purpose, an FRP cell server can be attached to every switching element by dedicating one (or several) inputs and outputs (see figure 4).

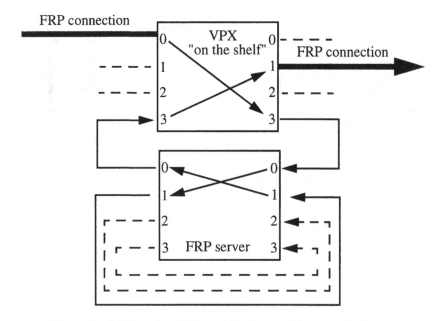

**Figure 4**   Addition of an FRP capability to an existing network element

A particular connection (represented as a bold line) is arriving on input multiplex 0 (IM#0) and departing on output multiplex 1(OM#1) of a switching element bought "on the shelf" in 1993 when no statistical capability is implemented.

The ATM service provider now wants to adapt the cell rate allocation of this connection by using FRP/DT or FRP/ABR cells. He can do that by modifying the route so as to include the server - just by changing the connection entry in the switch translation table. There is no need to identify FRP cells in the switching element : data cells will pass as well through the FRP server.

Let us follow the arrows on the figure to see the new connection route : every cell is now directed to OM#3 (instead of OM#1), then to the FRP server on IM#0, then to server OM#1 (which mirrors switch OM#1) where FRP cells are specifically processed to perform bandwidth allocation, then to server IM#1, then to server OM#0 to join the switch on IM#3, and finally they are directed to switch OM#1.

Note that this arrangement is quite transparent to ATM connections not subject to FRPs. In particular any failure of that server would not alter these connections : this is particularly important when testing FRP equipment in ATM field trials actually under operations.

Link(s) between the switch and the FRP server will obviously become a bottleneck when a majority of ATM connections request an FRP service. In a next version of ATM switching elements, the FRP capability will migrate into the output ports - remaining fully compatible with the initial FRP service offer.

## 3.2 Upgrading the ATM UNI : the Mirror Source

In the next three years, Customer's Equipments (CEQs) may not implement FRPs ; we do not expect private routers to issue FRP cells towards the public ATM User Network Interface (UNI) until we have demonstrated its efficiency in field trials. However, FRPs can be used as "network-internal" traffic control procedures running within network edges. They would complement other traffic controls run between the customer equipment and the public network.

Agreement on a traffic control procedure between CEQ and public network through the ATM UNI is a subject of severe competition among industrials and seems to be a big issue. At least, two ways can be investigated : Generic Flow Control [23] and Q.2963 B-ISDN signalling [24].

A specific equipment can be introduced on the public network side of the ATM UNI to translate these control signals between the user and the public network into FRP signals run internally : this is the Mirror (traffic) Source (see figure 5):

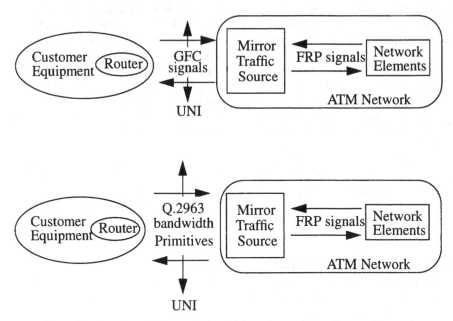

**Figure 5**   Mapping a UNI control into FRP used as an internal network control

## 4   WHO CAN BENEFIT FROM FAST RESERVATION PROTOCOLS ?

In general, users would like to perceive the allocation of a peak cell rate while paying only for network usage : they do not really care for bandwidth allocation pertaining during idle periods. Fast Reservation Protocols have been intended for that but they are efficient only with particular users.

These ATM users produce variable bit rate traffics and would like to be charged according to their actual activity. They request rates which are a small fraction of network element port rate - say, smaller than 1/30th. They specify Quality-of-Service requirements into the traffic contract - at least, regarding information loss or transfer delay or both.

They cannot predict accurately their activity but they are aware that a stepwise structure is embedded into the evolution in time of the generated cell rate. They accept to delineate it for the ATM layer by means of FRM cells.

They accept that the network makes use of this delineation to perform an access control. Depending upon the Fast Reservation Protocol to be used, the duration of traffic steps may have to be sufficiently large compared to the round trip delay along the connection - say, 10 times larger than the round-trip. As a counterpart, they are allocated the peak bandwidth they need nearly immediatly when they need it.

Here follow some examples of applications an ATM network can offer efficiently by means of FRP connections. We assume that FRP signals are conveyed in both directions through the user-network interface. Otherwise, a Mirror Traffic Source can be used at the network side - as previously shown.

We think that an FRP capability could be included first in the architecture of peripheral network elements (e.g. access multiplexers) and new servers designed for multimedia applications (e.g. home shopping, still picture banking and retrievals, video-on-demand, etc...) .

An ATM concentrator in the local loop can use FRPs to enable several users to share a high-speed long-distance (expensive) link accessing the public carrier transit network.

A third example consists in tailoring virtual private networks on top of an ATM network infrastructure. Following the idea of S. Walters [25], users purchase a set of point-to-point virtual path (VPCs) within which they define cross-connected virtual circuits (VCCs). The network operator allocates a peak bandwidth to every VPC and guarantees any specified QoS. We propose that VCCs within a VPC share (either statistically or not) the VPC peak cell rate by means of FRP/ DT cells. This proposal solves for the "output policing" operator's problem and provide the user with a VCC contention resolution algorithm.

Then, terminal equipment could implement an FRP capability. For example, a router in the customer equipment could use FRP/IT to delineate CS-PDUs to be transferred through the public ATM network.

## 4.1 A Multimedia Server

Distance-learning, video-conferencing with variable quality, oject visualization (even, "virtual reality" applications), image retrievals are business applications an ATM service provider could support with profit soon. In a near future, such residential communication services as video-on-demand and home-shopping will also take-off.

All these applications are multimedia : they combine text, audio, still images and full motion video with several picture qualities and bit rates. They are currently stand-alone but users might like to extend them to include distant multimedia servers.

These multimedia servers would be reached via ATM connections set-up in the public network. Each connection would support requests from the user to the server and multimedia data from the server to the user. At the server access to the ATM network, a typical multimedia connection offers a stepwise variable bit rate in time (see figure 6).

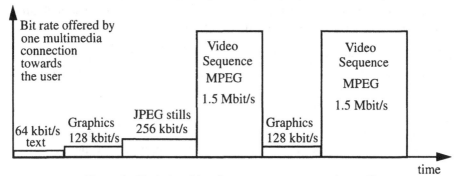

**Figure 6**   Typical multimedia server-to-user connection traffic

The pricing of multimedia connections can be much reduced if the server concentrates traffics at its output. FRP/DT is ideally suited to the concentration of this kind of traffics.

First, users are likely to accept to wait during a round-trip at the electrical propagation speed - say, less than 50 ms - before starting its new phase of activity. Moreover, step durations are in the range of several seconds which is fairly long compared to the round-trip delay. Eventually, successive peak cell rate values are below 1/30th the link rate which allows for a good multiplexing gainDuring the FRP/DT cell round-trip, the new peak cell rate is reserved into the ATM network switching elements ; then, the ATM network is able to offer a circuit-mode quality with low cell delay, low cell dispersion and negligible cell loss rate.

## 4.2  Sharing an High-Speed Access Line to ATM Network

Many potential users cannot afford a high-speed (155 Mbit/s) ATM multiplex to be connected to the ATM network - in particular when they need much smaller ATM service rates. Therefore, they could be interested in statistically sharing this high-speed line.

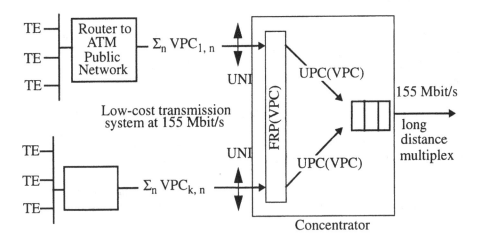

**Figure 7** Statistical sharing of a long-distance high-speed access line to transit network

Since they do not intend to communicate with each other, there is no need to use MAN capabilities ; a solution (see figure 7) based upon a concentrator/de-concentrator operated by the ATM service provider matches.

The concentrator should be installed in the vicinity of a group of users (e.g. , at the basement of a business block) so that users would have to pay only for a short-distance high-speed transmission link.

The ATM service provider allocates resources to VPCs while the VCC level is left to user internal multiplexing. For each VPC, a peak cell rate value is allocated and policed by the concentrator ; this value results from an in-call bandwidth negotiation with the router performed by means of FRP cells. When some VPC request is denied, the bandwidth increase can be either postponed until network agreement (FRP/DT) or cells will be discarded (FRP/IT).

The allocation of a peak cell rate guarantees the most demanding user's QoS.A user may segregate its traffics into VCCs but it has to control their Quality-of-Service. The network provides FRM capabilities to solve for contention between VCCs.

## 4.3  A Virtual Private Network service

If we consider the particular example of a firm trying to interconnect its facilities, private lines may not be an adequate solution as soon as the number of locations grows. Alternatively, a virtual private network (VPN) could be worth considering.

Steve Walters has proposed an interesting type of virtual private network using virtual circuit cross-connects and virtual path policing [25]. The virtual private network is composed of VPCs. These paths are allocated a peak bandwidth and their quality-of-service is guaranteed by the network.

The user predefines any number of cross-connected virtual circuits VCCs between any pairs of facilities. These circuits are supported by the virtual private network VPCs . The user is responsible for the quality-of-service of his VCCs.

The ATM service provider has to police the peak bandwidth allocated to each VPC. But, a VPC may start at the output of a network element, agregating the traffic of several incoming VPCs. Walters proposes to solve the problem by performing a policing function in network element output port - so-called "output policing" (see figure 8). The implementation is not trivial because cells may arrive in an output port n times faster than the output port speed.

Moreover, if "output policing" were achieved, it would cause uncontrolled cell loss at the VCC level.

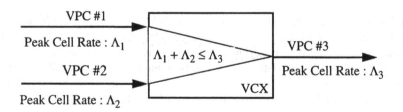

**Figure 8**   The VPC "output policing" problem

Indeed, Walters does not provide the user with any kind of protocol he could run to control VPC contention within his VPN. A VPC departing from a cross-connect exceeds its allocation when it supports too many VCCs. The excess traffic will be discarded by policing to guarantee the other VPC QoS. Alternatively, the number and traffic activity of VCCs supported by a VPC could be controlled from the user's side to avoid VPC overflow.

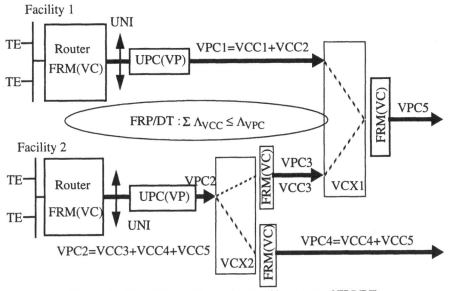

**Figure 9** Virtual Private Network control by means of FRP/DT

The FRP/DT protocol can be run throughout the ATM network to allocate bandwidth to every VCC within the sucessive VPCs up to their respective VPC peak cell rates - (see figure 9).

# 5  CONCLUSION

Initial ATM services are based upon permanent connections. To be viable, ATM service providers are bound to achieve some statistical sharing of the transmission capacity they lease from infrastructure network providers.

Switching is a classical way to achieve some statistical sharing of the network resources  at the call level. Indeed, a switched connection is set-up only when necessary and break when no longer needed. Many efforts are made within the standardizing bodies to finalize signalling systems ; many others will have to be made to implement them.

However, switching reacts at the call level scale of time - say, several hundreds of ms are required to accept a candidate call. It achieves a good statistical gain in STM environment where randomness lies only at the call level and user accepts delay up to some extent before set-up.

In many situations, switching is too slow to adapt the network resource allocation to real-time traffic profiles of broadband traffics. Alternatively, the user of an ATM connection could be allowed to purchase a temporary increase of network resource that would be allocated within the smallest amount of time - even reducing to the electrical propagation delay along the connection route. This is the Fast Resource Management concept.

We have shown several examples of early applications where it can be viewed as resource switching for permanent connections. But it could be used as well to allocate resource to switched connections. In both cases, it allows for usage-based pricing of ATM connections.

Among Fast Resource Management protocols, some Fast Reservation Protocols can be run on top of ATM networks without replacing existing equipments.

Compared to permanent connections, switched connections have always represented some «risk» for a public carrier because it has a few seconds to trade-off between network efficiency and quality-of-service at call set-up.

These «risks» are re-inforced within a broadband network. Indeed, the admission control is sensitive to traffic characteristics. What we know about broadband traffics and user behaviours is very vague and it may take time to capture them.

Users badly accept to be denied admission but this is a practical and safe way to guarantee QoS to both newly admitted connection and other connections already set-up. Fast Reservation Protocols extend this concept at a burst level within the real-time traffic supported by ATM connections.

# REFERENCES

[1]  CIDON, I., GOPAL, I., SEGALL, A., "Fast Connection Establishment in High-speed Networks," report RC 15515, IBM T. J. Watson Research Center, February 1990.

[2]  ITU Recommendation I.371, "Traffic Control and Congestion Control in B-ISDN," Section 3.2.6, Geneva, March 1994.

[3]  ITU Recommendation I.361, "B-ISDN ATM Layer Specification," Section 2.2.4, Geneva, June 1993.

[4]  AMSTUTZ, S.R., "Burst Switching - An Introduction," IEEE Communication Magazine, November 1983.

[5]  HUI, J.Y., "Resource Allocation for Broadband Networks," IEEE Journal on Selected Areas in Communications, Vol. 6, December 1988.

[6] OHNISHI, H., OKADA, T., NOGUCHI, K., "Flow Control Schemes and Delay/Tradeoff in ATM Networks," IEEE Journal on Selected Areas in Communications, Vol. 6, December 1988.

[7] GIRARD, A., BOYER, P.E., COUDREUSE, J-P., "Réservation de Débits dans un Réseau de Paquets Asynchrones," French Patent 89-02073, February 1989.

[8] BOYER, P.E., "A Congestion Control for the ATM," 7th ITC Seminar, Morristown, NJ, USA, October 1990.

[9] DOSHI, B.T., DRAVIDA, S., "Congestion Control for Bursty Data in High-speed Wide Area Packet Networks : In-call Parameter Negotiations," 7th ITC Seminar. Morristown, NJ, USA. October 1990.

[10] TURNER, J. S., "A Proposed Bandwidth Management and Congestion Control Scheme for Multicast ATM Networks," Technical Report WU-CCRC-91-1, Washington University, May 1991.

[11] BOYER, P.E., TRANCHIER, D.P., "A Reservation Principle with Applications to the ATM Traffic Control," Computer Networks and ISDN Systems, Vol. 24, May 1992.

[12] SUZUKI, H., TOBAGI, F.A., "Fast Bandwidth Reservation Scheme with Multi-link and Multi-path Routing in ATM Networks," INFOCOM'92, Firenze, April 1992.

[13] OHTA, S., SATO, K-I., "Dynamic Bandwidth Control of the Virtual Path in an Asynchronous Transfer Mode Network," IEEE Transactions on Communications, Vol. 40, July 1992.

[14] DRAVIDA, S., DOSHI, B., HARSHAVARDHANA, P., "Performance and Roles of Bandwidth and Buffer Reservation Schemes in High-speed Networks," ITC'13, June 1994.

[15] BOYER J., "Statistical Multiplexing of Data ATM Connections," 2nd International Conference on Telecommunication Systems Modelling and Analysis, Nashville, TN, USA, March 1994.

[16] ENSSLE, J., BRIEM, U., KRONER, H., "Performance Analysis of Fast Reservation Protocols", IFIP TC6 2nd Workshop on Performance Modelling and Evaluation of ATM Networks, July 4-7, 1994, Bradford, UK.

[17] JAIN, R., "Myths about Congestion Management in High-Speed Networks," 7th ITC Seminar, Morristown, NJ, USA, October 1990.

[18] The ATM Forum, "ATM User-Network Interface Specification", Version 3.0, September 1993.

[19] ITU-TSS, Study Group 13, "The Fast Reservation Protocol", Delayed Contribution D71, July 1993.

[20]  CHARNY, A., CLARK, D., JAIN, R., "Congestion Control with Explicit Rate Indication", ATM Forum Document 94-0692, July 1994.

[21]  ADAMS, J., SMITH, A., "The Dynamic Bandwidth Controller : A Proposal for the ABR Servicel", ATM Forum Document 94-0555R1, June 1994.

[22] AOYAMA, T., TOKIZAWA, I.., SATO, K-I.., "An ATM VP-based Broadband Network : its Technologies and Application to Multimedia Services," ISS'92, October 1992.

[23] ITU-TSS, Study Group 13, "ATM Aspects," Meeting Report SWP2/1, March 1994.

[24] ITU-TSS Study Group 11, Draft Recommendation Q.2963 "B-ISDN DSS2 : Connection Modification", Edinburgh (UK), June 1994.

[25] WALTERS, S. M., "A New Direction for Broadband ISDN," IEEE Communications Magazine, September 1991.

# 5

# FLOW CONTROL AND SWITCHING STRATEGY FOR PREVENTING CONGESTION IN MULTISTAGE NETWORKS

A. Pombortsis, I. Vlahavas

*Department of Informatics, Aristotle University of Thessaloniki,*
*54006, Thessaloniki, Greece*
*E-mail: pomportsis/vlahavas@olymp.ccf.auth.gr*

## ABSTRACT

Multistage Interconnection Networks are used in parallel computer applications as well as in new, high performance packet switch architecture for communication systems. In these networks the presence of unbalanced traffic loads creates significant performance problems. The main goal of this paper is to propose: (a) a flow control scheme, which permits the characterization of the traffic distribution, and (b) switching strategies, at the packet level, for controlling this performance degradation. We also present a switch model, in order to support the suggested solutions. The proposed solutions can be implemented with minimal additional logic in the switch design. Simulation results are presented to test the effectiveness of the proposed approach.

## 1 INTRODUCTION

Self-routing, packet-switched Multistage Interconnection Networks (MINs) have been extensively used in parallel architecture for processor-to-processor or memory interconnections [1], [2]. Recently, a significant amount of interest has been generated in applying MINs to the design of the next-generation communication switching systems [3], [4].

Multistage Interconnection Networks in which a unique path exists between any input and output port, belong to the class of Banyan networks. The Banyan network was first proposed in [5] for multiprocessor systems. The basic component of an MIN is the Switching Element (SE) which is a 2x2 crossbar switch. An $N \times N$ MIN is constructed by arranging the SEs into

$n = \log_2 N$ stages, where each stage is consisting of an array of $N/2$ SEs. The linkages between various stages are assigned so that a one-to-one path can be established from any of the input ports to any of the output ports. An MIN can be unbuffered or buffered. Unbuffered MINs have buffers only at network inputs and outputs, while in buffered MIN buffers are embedded in some point in every SE. In [6] it has been shown that buffering of packets at the output ports of the switch is more efficient, in terms of network throughput, than buffering at the inputs. Though the buffered MINs are more complex than their unbuffered counterparts, they can provide higher throughput and are more attractive for the implementation of interconnection networks with high performance requirements. In the literature, several publications on the performance analysis of MINs of both types can be found [7], [8], [9], [10], [11], [12].

Generally the MINs have many advantages such as modular construction, hardware complexity proportional to $\mathbf{O}$ ($N$ $\log_2 N$ ), and distributed routing which allows the packet routing to be performed at a very high speed.

## Statement of the Problem and Solution Approaches

Packets passing into an input of an MIN may content with packets from other inputs for internal network links. This leads to blocking of packets and reduced network throughput. Although congestion in MINs is typically minimal if traffic patterns are perfectly uniform, the presence of unbalanced traffic loads, in which packets tend to follow fixed paths through the network creates significant performance problems. More precisely the MINs suffer from the serious problem of rapid and significant performance degradation when there is heavy traffic to one or a set of output ports. This traffic pattern is usually referred to as hot spot traffic and results in the so called tree saturation phenomenon [13]. The name arises from the fact that, when there is a hot output, after a short time period the entire fan-in tree, which leads to this hot output, will consist of full queues thus blocking both hot spot and regular packets alike. This traffic pattern may, for example, reflect the situation in a multiprocessor system when concurrent requests (packets) are submitted to the same shared variable or when, in a communication system, there are many outputs for data channels and one output for a dedicated video channel. It is interesting to note that the tree saturation can still occur (in short term) even under uniform traffic pattern [4].

Several methods have been proposed to solve the problem of tree saturation. In parallel processing systems the use of hardware combining (pairwise and three-way) as well the software combining, and memory scrambling have been proposed [13], [14], [15]. Some other solutions are based on the implementation of switching strategies into the switching elements, in order to control the packet flow in the network [16], [17]. The use of cascading and vertical stacking of multiple MINs has also been presented [18], [19]. Also in communication systems the use of a Distribution Network (DN), which is placed before the MIN and evenly distributes packets across all its outputs, has been proposed [4].

However, these methods are not generally applicable, since the tree saturation may arise from many reasons [14]. Also, in some cases the hardware augmentation required for such solutions is extremely expensive, thus weaken the basic advantage of the MINs, over other Interconnection Networks, which is the cost.

## Overview of the paper

The purpose of this paper is to propose a new approach which is based on a combination of a flow control scheme with switching strategies, and to present a switch description, in order to support the proposed approach. The flow control scheme is implemented by introducing a feedback mechanism in MINs. Each SE in a given stage i, can receive feedback information from SEs belonging to stages (i+1) through n (where $n=\log_2 N$ ), depending on the feedback policy. The feedback information permits the characterization of each packet as "hot" or "cool". Based on this characterization the combinations of two policies (that give priority to "hot" and to "cool" packets respectively) with two switch strategies ("queued" and "discard"), are studied. Simulation results of the proposed policies are provided.

The remaining sections are organized as follows: the principles of operation are briefly described in section II. This section is followed by the proposed flow control procedure (section III). Simulation results and discussion are in section IV. Finally, section V presents the conclusions of this research and describes future directions.

## 2  MODELING

The analysis is based on the following assumptions:
1.  A buffered, binary routing, packet-switched Multistage Interconnection
    Network is employed. Since the characteristics in terms of operation
    and performance, are similar for all different MIN topologies, we
    consider here an Omega MIN.
2.  The basic building blocks are 2x2 Switching Elements with output
    buffering.
3.  The MIN connects $N$ inputs (processors, with input buffer controller)
    to $N$ outputs.
4.  Each input "issues" $r$ packets to the network outputs per network cycle
    (where $0 \le r \le 1$). Among those packets, $h$ percent of the packets are
    hot spot packets directed to hot output.
5.  All packets are of fixed length.
6.  The network operates in a synchronous mode (the switch operation is
    slotted and the packets are transmitted only at the beginning of a time
    slot).
7.  All inputs behave in a stochastically identical way and the input arrival
    processes are independent.

## 3  THE FLOW CONTROL PROCEDURE AND SWITCHING STRATEGIES

The main function of flow control in packet networks is the prevention of
performance degradation due to congestion. The proposed here two flow
control procedures (priority and selection) aim to detect the possibility of
tree saturation, as soon as possible, and to prevent the creation of saturated
trees using proper switching strategies and utilizing the knowledge about
traffic load distribution.

A very interesting approach to the problem of preventing the tree saturation
phenomenon, by using feedback mechanisms, has been recently presented in
the literature [20]. The feedback scheme that has been used in [20] was very
simple, since it involved feedback information (the size of output queues)
only from the output of the network. In this paper we extend the feedback
mechanism by providing feedback information in each SE within the
network. The SE's decision whether to forward or to block (or reject) an
incoming packet is made according to local feedback information.

## 3.1   The feedback Schemes

The proposed feedback mechanism is based on the queue sizes of each SE, and works as follows: A threshold value (THV) is assigned to each SE of the network. If the size of a queue exceeds this certain THV or falls below this THV, the queue is assumed to be hot or cold respectively.

In order to implement the feedback policies each SE sends feedback information to the two SEs of the previous stages with which it is connected. The transmitted feedback information is organized in form of two tables (the Control Tables), corresponding to the two queues of a given SE. Each table has $2^k$ rows and k columns. Each row of the table corresponds to every possible destination of a given packet from the SE under consideration. The index k represents the number of controlled stages and can be in the range of 1 to $\log_2 N$ (i.e. the number of stages), depending on the feedback policy. Each cell of the table represents the state of a queue and may be "one" (the queue size exceeds THV) or "zero" (the queue size is smaller or equal than THV). At this point, it is worth noting that the feedback information, for k controlled stages, actually consists of only $2*[2^{(k-1)}-1]$ bits. From these bits the corresponding Control Table can be constructed by forming two binary trees which have as roots the two most significant bits of the transmitted information.

With respect to the feedback policy; (a) the feedback information is propagated from stage to stage (from the outputs to the inputs of the network), (b) consists of $2*[2^{(k-1)}-1]$ bits, where the first two most significant bits represent the status of the sending SE and the rest bits represent the status of the SEs of the next (k-1) stages, which are connected with the particular SE, according to the network topology. Obviously, these later bits are obtained from the two corresponding control tables of the SE under consideration. Thus, the Control Table represents the current status of the network paths from the given SE's port to the network outputs.

## 3.2  Example

In order to clarify the above issues we present the following example. Given a Multistage Interconnection Network of size $N \times N$, the n stages are numbered from 0 to (n-1), starting with the leftmost stage. Input links connected to SEs in a stage are numbered from 0 to (N-1) starting with the topmost input. If the SEs in every stage are numbered from 0 to N/2-1, then $SE_{si}$ represents the SE in row i and in column (stage) s, while $Fd_{si}$ represents the feedback information that $SE_{si}$ sends back to other SEs. Figure 1 shows a graphical example of an MIN with $N = 16$ and the flow of feedback information in a segment of it .

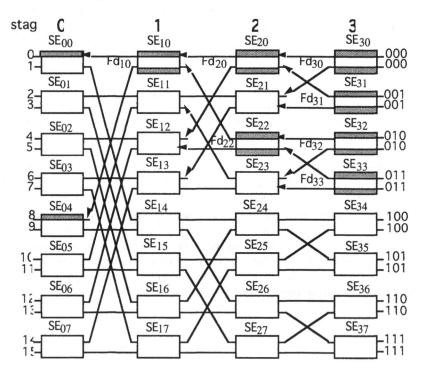

**Figure  1**      Flow of feedback information in a MIN of size 16x16

Let $Q^0{}_{si}$ , $Q^1{}_{si}$ and $CT^0{}_{si}$ , $CT^1{}_{si}$ represent the state (hot or cold) of the queues (0 for the upper queue and 1 for the lower queue) and the two control tables of the $SE_{si}$, respectively.

Considering the MIN of figure 1. In the case of one controlled stage (k=1), the operation of $SE_{10}$ includes the following three steps:

1. Receives the feedback information:

$$Fd_{20}+Fd_{22} = [Q^0{}_{20}+Q^1{}_{20}]+[Q^0{}_{22}+Q^1{}_{22}]$$

2. Constructs the following two control tables:

| | $CT^0{}_{10}$ |
|---|---|
| 0 | $Q^0{}_{20}$ |
| 1 | $Q^1{}_{20}$ |

| | $CT^1{}_{10}$ |
|---|---|
| 0 | $Q^0{}_{20}$ |
| 1 | $Q^1{}_{20}$ |

3. Sends back to the $SE_{00}$ and $SE_{04}$ the feedback information:

$$Fd_{10} = [Q^0{}_{10}+Q^1{}_{10}].$$

Considering the same MIN and Switching Element, but with two controlled stages (k=2) we have that the $SE_{10}$:

1. Receives the feedback information:

$$Fd_{20} +F\ d_{22} = \big[[Q^0{}_{20} + Q^1{}_{20}] + [Q^0{}_{30} + Q^1{}_{30}] + [Q^0{}_{31} + Q^1{}_{31}]\big] +$$
$$+ \big[[Q^0{}_{22} + Q^1{}_{22}] +[ Q^0{}_{32} + Q^1{}_{32}] + [Q^0{}_{33} + Q^1{}_{33}]\big]$$

2. Constructs the following two control tables:

| | $CT^0{}_{10}$ | |
|---|---|---|
| 00 | $Q^0{}_{20}$ | $Q^0{}_{30}$ |
| 01 | $Q^0{}_{20}$ | $Q^1{}_{30}$ |
| 10 | $Q^1{}_{20}$ | $Q^0{}_{31}$ |
| 11 | $Q^1{}_{20}$ | $Q^1{}_{31}$ |

| | $CT^1{}_{10}$ | |
|---|---|---|
| 00 | $Q^0{}_{22}$ | $Q^0{}_{32}$ |
| 01 | $Q^0{}_{22}$ | $Q^1{}_{32}$ |
| 10 | $Q^1{}_{22}$ | $Q^0{}_{33}$ |
| 11 | $Q^1{}_{22}$ | $Q^1{}_{33}$ |

3. Sends back to the $SE_{00}$ and $SE_{04}$, the feedback information:

$$Fd_{10} = [Q^0{}_{10}+Q^1{}_{10}] + [Q^0{}_{20}+Q^1{}_{20}] +[Q^0{}_{22}+Q^1{}_{22}].$$

## 3.3   The Priority Algorithm

Based on the "Control Table" values and by decoding the routing tag, each packet can be characterized as "hot" or "cool" taking into account the current status of the network paths. One more bit (the "priority flag" P) can be added to the head field of the data packets. Depending on the priority policy, the two kind of packets, ("hot" and "cool") are labeled as high (P=1) or low priority packets. In case of packet contention at the SEs (i.e when packets from different switches attempt to advance simultaneously to the same output port of a switch in the next stage) we have that:

(i)    in the "queued" switching strategy, and in case of contention, the high priority packet prevails over low priority packet.

(ii)   in addition, in the "discard" strategy the low priority packets are discarded and the input retransmits the discarded packet (i.e. an Accept, Otherwise Reject (AOR) strategy [21)] is used.

Within each priority, the service priority is FIFO and in case of congestion between packets of the same priority one of them is randomly selected.

The complete listing of the algorithm is given in a Pascal like in the Appendix.

## 3.4   The Selection Algorithm

Using the above described feedback information, each SE submits packets to the next stage, according to the two Control Tables corresponding to its queues. Every SE forwards to the next stage not the packets that are found in the head of the queues (FIFO strategy), but the packets that are destined to non-hot output ports, (i.e. to ports which are represented by "zero" in the Control Tables). In the case where no such packets exist, the packets at the head of the queue are selected for forwarding. The complete listing of the algorithm is also given in a Pascal-like in the Appendix.

# 4 SWITCHING ELEMENT IMPLEMENTATION AND OPERATION

The Switching Elements are required to perform the routing of the packets. A block diagram of of a 2x2 SE that can fulfill the requirements of the proposed solutions is shown in Figure 2. As mentioned in the previous section, the feedback mechanism used by every SE in a network cycle consists of three stages: (i) Reception (ii) Construction and (iii) Transmission, i.e. in every network cycle, each SE receives feedback information, constructs the two Control Tables and sends back feedback information. The first and third stages are overlapped.

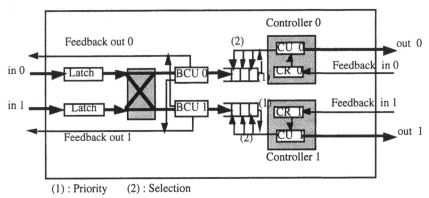

(1) : Priority     (2) : Selection

**Figure 2**     Block diagram of a 2x2 switching element with priority/selection capabilities.

To implement the flow control procedure proposed in this paper, each queue in the SE is modified so that the destination address either i) of the packet held on the top of this queue (priority algorithm) or ii) of every packet held in the queue (selection algorithm), can be compared by a Control Unit (CU), with the content of the associated Control Register (CR).

The control register contains the control table and is $2*(2^k-1)$ bits wide, where k is the number of controlled stages. The control register and the control unit constitute the controller associated with each queue. An incoming packet is buffered by a Buffer Control Unit (BCU) at the end of one of two buffer queues, depending on the destination output port. Furthermore, the BCU compares the size of the queue with the THV and sends back the appropriate feedback information.

Packets received from the preceding stage and feedback information received from the following stage are stored in the queue and the control register respectively. At the same time the Control Unit either assigns priority values to the packets founded to the head of the queue (priority algorithm) or selects the appropriate packets, and forwards them to the next stage. Therefore, each cycle must be long enough to permit two transmissions between consecutive stages of the network and to carry out the necessary logic operations.

# 5   SIMULATION STUDY AND RESULTS

Because of the service policy introduced by the switching elements, an exact analysis of the network under study is untractable. Thus, this section presents results obtained from computer simulation of MINs of various sizes with a single hot output port and with buffer length equal to 4 packets, for various hot spot rates utilizing the proposed flow control procedure and the priority, selection strategies. In the simulations only the forward network is modeled since it is the network in which congestion occurs. If it is needed, the backward network can be taken into account by adding an appropriated number of network cycles to the network delays.

It is worth noting here that under nonuniform traffic patterns, we are interested not only for the mean values of the performance criteria but also in find the paths which have the worst congestion and investigate the effect of these paths on the "regular" paths and on the overall network performance. Thus the following network performance criteria were used:
(a)   The *general* $(E_G)$ network pass rates which is defined as the ratio of the total number of requests being serviced to the number of requests already "generated" (by processors or by network interfaces).
(b)   The *mean delay* $(D_m)$ which is defined as the ratio of the total time spend by the "generated" requests in the network to the number of requests already "generated". The delay is expressed in network cycles (n.c.).

The above criteria were also computed for packets which are addressing the hot output(s) and for regular (uniform) packets ($E_G{}^h$, $D_m{}^h$ and $E_G{}^u$, $D_m{}^u$ respectively). Also, in order to quantify the trade-off between the network throughput, which is expressed by the various network pass rates and the system response time, which is expressed by the delays, we use a combined performance criterion known as "power" (denoted by P), and it is defined as the ratio of the network pass rate to the network delay [22]. Thus, the performance criteria $P_G$, $P^h$, and $P^u$ were also computed.

Figure 3 presents the mean performance criteria of a MIN, without any control, for various hot spot rates (*h* %). In such a network as the hot spot rate increases, the mean network pass rate decreases and the mean delay increases. It is obvious that a higher hot spot rate results in more hot packets being using the network, and worsens the tree saturation. The higher the hot spot rate, the faster the network performance criteria drop off.

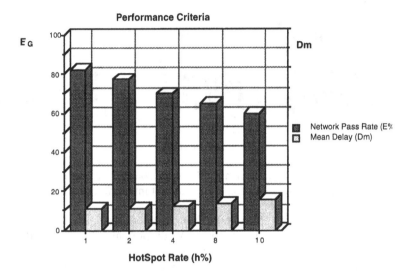

**Figure 3**    Mean performance criteria of a MIN, without any control, as a function of the hot spot rate (N=256).

Now let us examine the results of using the proposed solutions. Figure 4 compares the delays of hot/ regular packets and the mean network delay, for various hot spot rates, a) when priority is given to the "hot" packets, b) when priority is given to the "regular" packets, c) when the "hot" packets are discarded and d) when select the appropriate packet. It can be seen that under the discard strategy and when the hot packets are discarded (i.e priority is given to the "regular" packets), the delay of regular packets and the mean network delay are not very sensitive to the existence of unbalance traffic loads. Thus, the main objective which is to minimize the performance degradation of packets which do not participate in the hot spot activity, has been fulfilled.

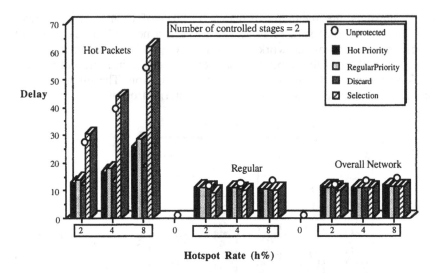

**Figure 4**      Delays comparison in an MIN with respect to the switching strategies as a function of the hot spot intensities, ($N$ =256).

The reason is as follows. Since the "hot" packets are discarded the tree saturation does not occur and the uniform component of traffic undergoes no degradation. In addition, the mean network delay is improved because it is dominated by the good delays of the regular traffic. Also, the discard strategy improves the delay of hot packets. Simulations show that the discard of regular packets does not help, since the network "overall" operation is dominated by the saturated tree(s) which leads to the hot output(s). Thus, if the main interest is in the network delays the discard strategy with priorities to the regular packets must be used.

Generally, different applications generate different communication patterns, and may have different performance requirements. Notice that the proposed solution does not require any *apriori* knowledge of traffic distribution. Since the answer how much one "hates" delay versus how much one "loves" throughput [22], depends primary on application, we also present various results using the "power" as performance criterion.

Figure 5 complements the previous results by plotting the throughput-delay characteristics (i.e. the power). Comparing the results from this figure, we notice that the assignment of priority to the "hot" packets improves the power of these packets and the overall power of the network. In this case the hot output is kept as busy as possible and the hot spot tree may clear earlier.

In contrast, by discarding the "hot" packets the power of the network is decreased. This is because the discarded packets wasting cycles and they must be retransmitted again.

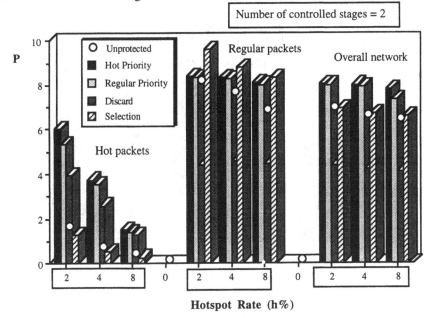

**Figure 5**     Throughput-delay characteristics (P) in an MIN with respect to the switching strategies, as a function of the hot spot intensities.

Finally, figure 6 compares the values of the power for a hot-spot intensity of 4% with respect to the number of controlled stages in the feedback schemes. Simulation results suggest that the control of two or three network stages is sufficient to improve the network operation.

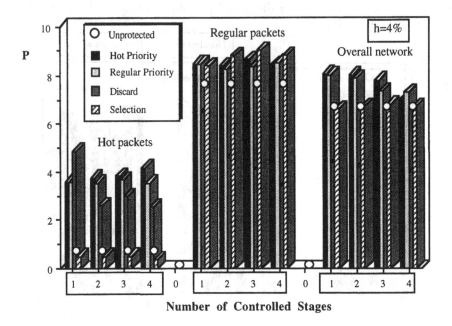

**Figure 6**    Variation of the network power with respect to the number of
control stages, (N=256, $h$ =4%).

We also simulated the network performance with multiple (2 and 3) hot
outputs. The results presented the same tendency. This means that the
method is also applicable in cases where multiple hot spots exist in the traffic
distributions.

# 6   CONCLUSIONS

In this paper, we presented a novel solution to the problem of performance
degradation of packet-switched Multistage Interconnection Networks, with
binary routing, under unbalanced traffic loads. The solution is based on a
combination of a flow control scheme with two priority policies and
switching strategies and permits the characterization of traffic distribution
within the network. Thus, it is possible to detect congestion and to restrict
packets that would aggravate this congestion. The proposed solution can be
applied both in parallel processing systems and telecommunication networks,
and generally in networks supporting traffic which is a fixed mix of two

types of packets or in networks serving mix traffic with different performance requirements. Extensive simulation results under various set-ups are provided. It is found from the results that the proposed solution offers a significant performance improvement, with an acceptable hardware complexity.

# REFERENCES

[1]  C. Wu and T Feng, "Interconnection Networks for parallel and distributed processing", IEEE Computer Society Press, (USA 1984).

[2]  H.J Siegel, "Interconnection Networks for Large-Scale Parallel Processing: Theory and Case Studies", Lexington Books, Lexington, (USA 1985).

[3]  G.E Daddis and H.C. Torng, "A Taxonomy of Broadband Integrated Switching Architectures", IEEE Communications Magazine, Vol. 27, No. 5, (May 1989), pp 32-41.

[4]  J.S Turner, "Design of a Broadcast Packet Switching Networks", IEEE Trans. Commun., Vol. 36, No 6, (June 1988), pp 734-743.

[5]  L.R. Goke and G.J. Lipovski, "Banyan network for partitioning multiprocessor systems", in Proc. 1st Annu. Symp. Comput. Architecture , 1973, pp.21-30.

[6]  M. Karol, M.Hluchyj, and S. Morgan, "Input versus output queueing on a space-division packet switch", IEEE Trans. Commun., Vol.35, No.12, (Dec.1987), pp.1347-1356.

[7]  J.H. Patel, "Performance of processor-memory interconnections for multiprocessor", IEEE Trans. Comput., Vol. C-30, No 10 (Oct.1981) pp 771-780.

[8]  D.M Dias and J.R. Jump, "Analysis and Simulation of Buffered Delta Networks", IEEE Trans. Comput., Vol. 30, No. 4 (Apr.1981), pp 273-283.

[9]  Y.C. Jenq, "Performance analysis of a packet switch based on single-buffered Banyan network", IEEE J. Select. Areas Commun., SAC-1, (1983), pp 1014-1021.

[10] C.P. Kruskal and M. Snir, "The performance of multistage interconnection networks for multiprocessors", IEEE Trans. Comput., Vol. 32, No. 12, (Dec. 1983), pp 1091-1098.

[11] C.P. Kruskal and M. Snir, "The distribution of waiting times in clocked multistage interconnection networks", IEEE Trans. Comput. Vol. 37, No. 12, (Dec. 1988), pp 1337-1352.

[12] T.H. Theimer, E.P. Rathgeb and M.N. Huber, "Performance Analysis of Buffered Banyan Networks", IEEE Trans. Comput., Vol. 39, No. 2, (Feb.1991), pp 269-277.

[13] G.F. Pfister and V.A. Norton, "Hot Spot Contention and Combining in Multistage Interconnection Network", IEEE Trans. Comput., Vol. 34, No. 10, (Nov. 1985), pp 934-948.

[14] P.C. Yew, N.F. Feng and D. Lawrie, "Distributing Hot-Spot addressing in large-scale Multiprocessors", Proc. of the 1986 International Conf. on Parallel Processing, (1986), pp 51-58.

[15] A. Norton and E. Melton, "A Class of Boolean Linear Transformation for Conflict-Free Power-of-Two Access", Proc. of the 1987 International Conf. on Parallel Processing,, (1987), pp 247-254.

[16] N. Tzeng, "Design of a Novel Combining Structure for Shared-Memory Multiprocessors", Proc. of the 1989 International Conf. on Parallel Processing,, (1989), Vol.I, pp1-8.

[17] D. Dias and M. Kumar, "Preventing Congestion in Multistage Networks in the Presence of Hotspots", Proc. of the 1989 International Conf. on Parallel Processing,, (1989), Vol.I, pp 9-13.

[18] S. Chalasani and A. Varma, "Analysis and Simulation of Multistage Interconnection Networks under Non-Uniform Traffic", Proc. of the PARBASE-90 , (1990), pp 258-265.

[19] C. Lea and D. Shyy, "Tradeoff of Horizantal Decomposition Versus Vertical Stacking in Rearrangeable Nonblocking Networks", IEEE Trans. Commun. Vol. 39, No. 6, (June 1991), pp 899-914.

[20] S.L. Scott and G.S. Sohi, "The use of feedback in multiprocessors and its application to tree saturation control", IEEE Trans. on Parallel and Distributed Systems, Vol. 1, No. 4, (Oct. 1990), pp 385-398.

[21] J-F. Chang and C-S. Wu, "The Effect of Prioritization on the Behaviour of a Concentrator Under an Accept, Otherwise Reject Strategy", IEEE Trans. Commun., Vol.38, No. 7, July 1990.

[22] L. Kleinrock, "Distributed Systems", Communications of the ACM, Vol. 28, No. 11, (Nov. 1985), pp 1200-1213.

# APPENDIX

A complete listing of the proposed flow control procedures is presented in a Pascal like form.

## *The Priority algorithm:*

**Begin**
    {Initialization}

        packet:=Pop(Queue);
        DAS:=Eval_DAS(packet);
        **case** policy **of**
        Priority_on_Hot:     *if* CT[DAS]=1          *then*
    packet.Priority:=High
                                                *else*     packet.Priority:=Low;
            Priority_on_Regular:           *if* CT[DAS]=0   *then*
    packet.Priority:=High
                                                *else*     packet.Priority:=Low
        **end of case;**

        **case** strategy **of**
        Queued:    Forward(packet);
        Discard:    *if*  packet.Priority=High *then*   Forward(packet)
                                                *else*     Discard(packet)
        **end of case;**

shift(Queue)
**End.**

## *The Selection algorithm:*

**Begin**
    {Initialization}

    Found:=false;
    p:=1;

    {Selection Loop}
    while (p<QP) and not Found do
        **Begin**

```
            packet:=Pop(Queue);
            DAS:=Eval_DAS(packet);
            if CT[DAS]=0  then   Found:=true
                          else  Begin
                                Rotate(Queue);
                                p:=p+1
                                End
        End; {end of selection loop}

        if Found then  packet:=Pop(Queue);
        Shift(Queue);
        QP:=QP-1
End.
```

Where:

| | |
|---|---|
| Queue[i], $0 \leq i \leq QL$ | Array of packets in a switch. |
| QL | Queue Length. The maximum number of packets that a queue can hold. |
| QP | Queue Pointer. The number of packets currently hold in a queue. |
| CT[i], $0 \leq i \leq 2^k$ | Control Table. Array of feedback information. Each cell of the array is K bits long and contains state information of the following K stages. |
| K | The number of controlled stages. K varies from 1 to $\log_2 N$. |
| P | Counter. |
| Found | Flag. |
| DAS | Destination Address Segment. Contains a segment of the destination address of a packet. It is K bits long, starting from the s+1 bit, where s is the stage number in which the particular SE belongs. |
| Pop(Q) | Function which returns the packet found on top of queue Q. |
| Eval_DAS(p) | Function that returns the destination address segment of packet p. |
| Rotate(Q) | Procedure that rotates one packet right the contents of queue Q. |
| Shift(Q) | Procedure that shifts one packet right the contents of queue Q. |
| Forward(p) | Procedure that forwards the packet p to the next stage. |
| Discard(p) | Procedure that discards the packet p. |

# 6

---

# PERFORMANCE MODELING AND CONTROL OF ATM NETWORKS

## Jon W. Mark

*Department of Electrical and Computer Engineering*
*University of Waterloo*
*Waterloo, Ontario, Canada N2L 3G1*

## ABSTRACT

Resource management issues are considered in two conceptual levels: (i) call admission and resource allocation for call connection establishment and (ii) traffic policing and service scheduling of in-progress calls. The functions in (i) are to be performed in the *Control Plane* while those in (ii) are to be enforced in the *Transport Plane*. We identify the functions associated with these two sets of resource management strategies, and incorporate these strategies into the end-to-end performance models.

## 1    PREAMBLE

Since the CCITT SG XVIII meeting in Seoul, Korea, in January 1988 to address issues towards meeting the broadband era, much work on the Asynchronous Transfer Mode (ATM) as a transport mechanism for broadband networking has been reported in the literature. The concern of this paper is a unified presentation of traffic management issues to shed an understanding of the operational features and characteristics of an ATM network from the points of view of both the user and the network premises. The role of *Resource Management* is to protect the network and the user to attain network performance objectives, and to enhance the utilization of the network resources. In general, we can view the resource management issues in terms of the time scale where management functions are performed. We distinguish *Resource Management* into *Network Management* and *Traffic Management*. These management functions are to be performed in different, but parallel, *planes*, as distinguished by the time scales. Traffic management functions are to be performed in the

*Transport Plane*, where the time scale corresponds to the cell level signaling speed. On the other hand, the network management functions are to be performed in the *Control Plane*, where the time scale is longer, say in the order of millisecond or larger.

Performance models for the traffic management layer provide a platform for the enforcement of control functions in the network management layer. We classify Call Admission Control (CAC) and resource allocation as network management functions. Traffic management functions include Usage Parameter Control (UPC), traffic shaping, service scheduling and buffer allocation. Service scheduling and buffer allocation address queue management in each of the backbone network nodes.

With the separation of call admission control and queue management functions into different time scale operations, their roles can be performed separately and independently, but are vertically coupled. In this work, we are concerned with traffic management issues, and with performance models that provide a framework for call admission control. As such, we focus attention on traffic management issues. The work presented in the sequel draws heavily upon the research results of the author and his collaborators, and those of other researchers reported in the literature.

# 2 CHARACTERISTICS OF AN ATM NETWORK

An ATM network is a mesh connection of multiplexers and switches in which switching and multiplexing are cell-oriented. A cell is a 53 byte data unit, consisting of 48 information bytes and 5 header bytes. The link transmission speed is nominally 155 Mb/s or 620 Mb/s. The mode of information transfer is connection-oriented. This means that a virtual connection between a source and a destination user must be established before the call can progress. The cell header contains two transport mechanisms: the Virtual Path Identifier (VPI) and the Virtual Channel Identifier (VCI). The established connections are then referred to as Virtual Path Connections (VPCs) and Virtual Channel Connections (VCCs). The reader is referred to [1] for terminologies, definitions, acronyms, etc., pertaining to ATM networking.

Connection establishment is a contractual agreement between the user and the network provider. As such, tangible factors pertaining to the user traffic and the required quality of service (QoS) parameters that are easily understood by both the user and the network provider must be defined. These factors are referred to as elements of a *traffic descriptor*. The traffic descriptor and the QoS parameters are intimately related. In essence, it is critically important to define a suitable traffic descriptor for which QoS such as cell loss rate, cell delay and cell delay variation can be explicitly specified. The agreed-upon traffic descriptor and the QoS to be provided under the agreement allow the network provider to exercise traffic management schemes to provide fair service and to protect network integrity.

## 2.1 Three Levels of Control

Resource management needs to be administered at three levels.

1. To facilitate connection-oriented service, call connections need to be established before a call is allowed to progress. To maintain network integrity and to provide agreed-upon quality of service, *call level admission control* must be enforced.

2. For in-progress calls, the cell-level traffic needs to be policed to control the amount of data allowed to enter the network by regulating violators. The policing of the individual traffic sources is a *cell level control*.

3. The usage parameter controlled traffic from the different users are multiplexed and scheduled for transmission onto the outgoing link to attain a prescribed level of quality of service. The multiplexing and scheduling of multiple traffic streams for transmission onto an outgoing link is a *burst level control*.

## *Call Level Admission Control*

Call admission control (CAC) is necessary for the network provider to achieve network performance objectives. Based on the traffic descriptor and the QoS requirements, the network has to decide whether or not a new call should be admitted. Once a call is admitted and a connection is established, it is incumbent upon the network provider to allocate sufficient network resources to satisfy the agreed upon QoS.

When a request for a new connection is initiated, a probing signal is dispatched along potential paths to establish a virtual connection. Each of the nodes through which the probing signal traverses needs to compute its residual re-source, defined as the total available nodal resource less the total resource already allocated to support in-progress calls. The amount of nodal resource allocated to support a given call shall be referred to as the *schedulable enve-lope*. That is, the user traffic which is allowed to enter the network must be within the limit specified by the schedulable envelope. We can infer that the backbone network defines and calculates the schedulable envelope, while the controller at the network access point enforces call admission control based on the limit specified by the schedulable envelope. We see that *scheduling* and *admission control* can be viewed as separable functions, which are linked to the extent that the scheduler provides the schedulable envelope and the admission controller observes this limit when admitting new calls.

## Cell Level Control

Some in-progress calls may attempt to send more than the agreed-upon traf-fic load, i.e., exceeding the limit specified by the schedulable envelope. If not curbed, the network can be driven into a chaotic state. It is therefore bene-ficial, and necessary, for the network provider to exercise cell-level control to *police* violators. Cell-level policing is to be exercised on a per-connection basis. Because of the high-speed, the transit delay between a boundary node and a potential bottleneck node can be very large, which can render reactive type of control for policing ineffective. Most traffic policers are of the preventive rate control variety, e.g., of a leaky bucket type [2].

## Burst Level Control

The link transmission speed of an ATM network is large compared to that of an individual user. To make efficient use of the network bandwidth, many individual user traffic are multiplexed at the network boundary for transmission over a single high-speed link. Through proper scheduling of service, statistical multiplexing can produce large gain. Scheduling and statistical multiplexing enhance link utilization to permit the asynchronous transfer mode to support a larger number of sources than the synchronous transfer mode. Thus, scheduling is the key to enhancing utilization while maintaining a certain degree of fairness among the participating users.

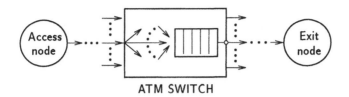

**Figure 1**  An End-to-End Network Partition

## 2.2  End-to-End Network Partition

In its most general form an ATM network provides multipoint-to-multipoint connections. The chaining of interior nodes to a pair of boundary nodes by transmission links provides many physical paths. In this paper, a physical path is referred to as an end-to-end network partition, which can support a large number of virtual connections. Let $\Omega_{np}$ be an end-to-end network partition connecting an access node to an exit node, as shown in Figure 1. In the context of Figure 1, an interior node has multiple input links but only one output link. The latter corresponds to the physical attributes of one output port of a multi-input multi-output, output-buffered ATM switch. This model of $\Omega_{np}$ ignores resource sharing among the output ports of an ATM switch. In our network partition, $\Omega_{np}$, every interior node is similarly structured; we use the term generic node to represent an interior node. With multi-input ports and a single output port, the generic node has an output buffer and a server in the form of the transmission link. The amount of nodal resource is represented by the 2-tuple $(B, c)$, where $B$ is the buffer space and $c$ is the transmission capacity. The traffic flows passing through the generic node all vie for a share of the available resource; hence the generic node performs statistical multiplexing and service scheduling.

In addition to supporting the feed-through traffic along the end-to-end virtual connections, each of the interior nodes in $\Omega_{np}$ also handles cross traffic, where a cross traffic is defined as any traffic flow that does not enter the next node in $\Omega_{np}$.

## 2.3   Separation of Traffic Management Functions

As mentioned earlier, call admission, traffic policing, service scheduling and queue management are resource management functions. For the network partition $\Omega_{np}$ stated above, the network access node oversees all four resource management functions, whereas an interior node only needs to perform service scheduling and queue management. In the sequel we will separately describe the performance modeling of the access node and the generic interior node. The access node multiplexes external arrivals, whereas the interior generic node multiplexes different traffic flows that are already in the backbone network. To facilitate performance modeling, we will next describe the traffic models for both external and internal traffic flows.

## 3   TRAFFIC MODEL

ATM traffic is expected to be heterogeneous, but there is no universally accepted traffic model to characterize the ATM traffic. In this work, we assume the external traffic streams arriving to the network access points to be *on/off* sources, with exponentially distributed *on* and *off* periods. The traffic parameters of each traffic type is to be described by the triple $(a, b, \gamma)$, where $a$ is the mean *on* period, $b$ is the mean *off* period, and $\gamma$ is the cell generation rate while in the *on* period.

## 3.1   Traffic Characteristics at the Access Node

In general, the aggregated input to the access node is a combination of $K$ traffic types, e.g., file transfer data, variable-bit rate coded video, voice, etc. Here, we consider these as $K$ types of *on/off* sources. Suppose type $k$ has $N_k$ sources ($k = 1, 2, .., K$). The total number of sources is then $N = \sum_{k=1}^{K} N_k$. Kosten [3] has presented a lucid study of the multiplexing of groups of multiple sources.

**Assumptions:**

1. The aggregate input consists of $K$ *on/off* types with different traffic parameters. A type $k$ traffic source has a parameter triple $(a_k, b_k, \gamma_k)$. The source inputs cells at the rate $\gamma_k$ while it is in the *on* period; it inputs

no cell while in the *off* period. The durations of the *on* and *off* states are exponentially distributed with means $a_k$ and $b_k$, respectively.

2. All input sources are independent of each other.

3. The sources share the output link of capacity $c$ and the buffer of size $B$ cells.

4. A given source can only be one of the $K$ types, so that $K \leq N$.

When $K$ is large, the queue management at the multiplexing buffer is quite complex. Lee and Mark [4] have presented an approach to characterize *on/off* sources using two *on/off* types to greatly facilitate the queueing analysis.

## 3.2 Traffic Characteristics at the Generic Node

Although the external arrival processes are *on/off* exponential, they may not be *on/off* or exponentially distributed once they enter the network. It is shown in [5] that a reference *on/off* traffic stream emerging from an output port of the generic node is approximately *on/off*. However, it is observed that (i) the durations of the *on* periods are lengthened, (ii) the *on/off* periods are no longer exponential, and (iii) the parameters $(a_k, b_k, \gamma_k)$ are modified.

Consider the $i$th node in the reference partition, $\Omega_{np}$. Let there be $N$ traffic flows, consisting of $K$ *on/off* types, that pass through the $i$th node. For simplicity, assume $N = K$, i.e., traffic flows of the same type are merged so that the resultant flows are all distinct (approximately) *on/off* types. The parameters of the $k$th traffic stream (type) at the $i$th node is described by the triple $(a_k^{(i)}, b_k^{(i)}, \gamma_k^{(i)})$. The available resource of the generic node is the buffer space and the link capacity. The problem of resource allocation is to partition and to allocate capacities to serve the $K$ traffic flows, subject to a satisfaction of the QoS requirements. That is, the problem is how to partition the capacity into regions, $\{R_1^i, R_2^i, ..., R_K^i\}$, in the $K$-dimensional space in which the performance criteria are satisfied. The $k$th region represents the schedulable envelope for the $k$th traffic flow and the union of the $K$ regions governs the global schedulable envelope of the $i$th node, which is also the limit for traffic admission into the network. Let $c_e^{(i,k)}$ be the effective capacity requirement of the $k$th traffic flow at node $i$ and $c^{(i)}$ be the total capacity at node $i$. It is required that

$$\sum_{k=1}^{K} R_k^{(i)} c_e^{(i,k)} < c^{(i)},$$
(6.1)

and

$$\hat{c}^{(i)} = c^{(i)} - \sum_{k=1}^{K} c_e^{(i,k)}$$
(6.2)

is the capacity available for admitting new calls. Let there be $I$ such interior nodes in partition $\Omega_{np}$. The schedulable envelope for admitting new calls into partition $\Omega_{np}$ is then

$$\hat{c}_{\min} = \min\{\hat{c}^{(i)}, \ i = 1, 2, ..., I\},$$
(6.3)

where $\hat{c}_{\min}$ is the schedulable envelope of the bottleneck node. The effective capacity, $c_e^{(i,k)}$, for handling the $k$th source corresponds to the sustainable cell rate (SCR) defined in [1].

To facilitate performance analysis, we make the following

**Assumptions**

1. The traffic streams are independent of one another.

2. The traffic streams share the output link of capacity $c$ and the buffer of size $B$ cells.

Note that for notational convenience, we use the same symbols to represent the characteristics at the boundary and at the interior nodes.

## 3.3   Performance Criterion

One performance criterion at both the access node and the generic interior node is cell loss rate due to insufficient buffer space. Let $Q(t)$ be the queue length at time $t$ under an infinite buffer assumption. If the probability of error is $P_e \ll 1$, then $P_e$ can be approximated as the tail of the queue length distribution [3]:

$$P_e \approx \lim_{t \to \infty} \Pr[Q(t) > B].$$
(6.4)

The key element in defining the schedulable envelope is to find an expression for $P_e$ in terms of the capacity regions $(R_1, R_2, ..., R_K)$. Even an expression were available, the computational complexity would increase as the number of flows, $K$, increases. Anick et al. [6] have found a simple solution for the case $K = 1$ and $N > 1$.

Suppose each of the individual external sources are controlled using a leaky bucket device before forwarding to the multiplexer. The leaky bucket capacity, as represent by the 2-tuple $(r_T^{(k)}, T_k)$, where $r_T^{(k)}$ is the token generation rate and $T_k$ is the token pool size, is a traffic descriptor element imposed by the network provider. However, contract negotiation should also reflect the characteristics of the source. The *effective bandwidth*, defined as the minimum channel capacity required to satisfy a prescribed cell loss rate [7, 8, 11], has also been used as an element of the traffic descriptor. The effective capacity requirement, $c_e^{(i,k)}$, mentioned above can be viewed as equivalent to the effective bandwidth. In this formulation, the nodal resource represented by the 2-tuple $(B^{(i)}, c^{(i)})$ specifies the limits for admission control, based on an effective bandwidth traffic descriptor and the associated QoS requirements.

A traffic stream passing through a node will experience transit delay (queueing delay, service time, processing time, etc.). This phenomenon is equivalent to low-pass filtering, which has the effect of smoothing out high frequency fluctuations. The departure process from the node can be thought of as produced by the convolution of the arrival process with the *impulse response* of the queueing model characterizing the node. Since convolution of two arbitrary functions results in a new function with width equal to the sum of the widths of the component functions, it follows that the queueing and service performed at the node has the effect of elongating the *on* periods of an arbitrary traffic stream. An interesting topic would be the search for an *impulse response* representation of a single-server queue.

Because of the "pulse" elongation effect, the end-to-end cell delay and cell delay variation (CDV) are also QoS parameters. The changes in cell delay and CDV are reflected by the changes in the triple $(a_k^{(i)}, b_k^{(i)}, \gamma_k^{(i)})$. The values of $(a_k^{(i+1)}, b_k^{(i+1)}, \gamma_k^{(i+1)})$ at the $(i+1)$st node can be expressed as those at the $i$th node. Let $j_k = 1$ (0) denote the *on (off)* state of the $k$th flow, $(1 \le k \le K^{(i)})$, $\underline{j} = \{j_1, ..., j_k, ..., j_{K^{(i)}}\}$ denote the state of the $K^{(i)}$ streams, and $\bar{q}_{\underline{j}}$ denote the average joint queue length. It is shown in [5] that the traffic parameters at successive nodes in $\Omega_{np}$ can be computed recursively:

$$a_k^{(i+1)} = a_k^{(i)} + \Delta a_k^{(i)}, \tag{6.5}$$

$$b_k^{(i+1)} = b_k^{(i)} + \Delta b_k^{(i)}, \tag{6.6}$$

$$\gamma_k^{(i+1)} = \gamma_k^{(i)} + \Delta \gamma_k^{(i)}, \tag{6.7}$$

where

$$\Delta a_k^{(i)} = [ \sum_{\underline{j}|j_k=1} \bar{q}_{\underline{j}}/\bar{\pi}_k^{(i)} - \sum_{\underline{j}|j_k=0} \bar{q}_{\underline{j}}/\pi_k^{(i)}]/c^{(i)}, \tag{6.8}$$

$$\Delta b_k^{(i)} = -\Delta a_k^{(i)}, \tag{6.9}$$

$$\Delta \gamma_k^{(i)} = -\frac{\Delta a_k^{(i)}}{a_k^{(i)} + \Delta a_k^{(i)}} \gamma_k^{(i)}, \tag{6.10}$$

and

$$\pi_k^{(i)} = \frac{\gamma_k^{(i)}}{a_k^{(i)} + b_k^{(i)}},$$

with $\bar{\pi}_k^{(i)} = 1 - \pi_k^{(i)}$, is the stationary probability of the $k$th process. The cumulative pulse elongation, as given by (6.8) to (6.10), permits the definition and caculation of the cell delay variation (CDV) tolerance.

# 4   ENFORCEMENT OF CONTROL AND SCHEDULING FUNCTIONS

As alluded to above, the boundary node, which multiplexes user traffic onto the backbone network, has the tasks of enforcing call admission control, traffic policing at the cell level, and service scheduling and queue management at the burst level. An interior node is concerned with service scheduling and queue management to calculate the schedulable envelopes for admission control. Thus, call admission control and scheduling functions can be separated. In this section we address the performance characteristics of the boundary node and a generic interior node separately.

## 4.1   Roles of the Boundary Node

From the ATM layer's perspective, the boundary node is implemented by a statistical multiplexer with proper policing mechanisms preceding it and proper traffic shaping mechanism following it. We assume the arrival traffic has been appropriately prepared by the ATM Adaptation Layer (AAL) and the arrivals are in a cell format. In this context, the policing mechanisms and the statistical multiplexer together form the network access node.

We assume that each source stream is policed by a leaky bucket rate controller. The key element is the rate at which tokens are generated. The token generation for a particular leaky bucket controller is a function of the effective capacity (sustainable cell rate) allocated for the particular source.

## Capacity Allocation at the Access Node

The resources at the access node is prescribed by the available buffer space, $B$, and the service capacity, $c$. Let $c_e^{(k)}$ be the effective capacity allocated to the $k$th stream. $c_e^{(k)}$ can be expressed as a function of the source characteristics $(a_k, b_k, \gamma_k)$, and the cell loss rate $P_e$. Using the cell loss criterion, the effective capacity required to serve the $k$th source is estimated by Gibbens and Hunt [7] as

$$c_e^{(k)} = \frac{a_k^{-1} + b_k^{-1} + \gamma_k + \sqrt{(a_k^{-1} + b_k^{-1} + \gamma_k \zeta)^2 - 4a_k^{-1}\gamma_k \zeta}}{2\zeta}, \quad (6.11)$$

where

$$\zeta = \frac{1}{B} \log \frac{1}{P_e}. \quad (6.12)$$

In [7], $c_e^{(k)}$ is called the effective bandwidth, which is a source specification. In our case, $c_e^{(k)}$ is a partition of the total capacity $c$, which the network provider allocates to support the $k$th traffic stream. The formula (6.11) given in [7] is an approximation by retaining only the term corresponding to the dominant negative eigenvalue in the power series spectral expansion of the queue length distribution [9, 10]. Expression (6.11) ignores the statistical multiplexing gain and contributions from non-dominant negative eigenvalues. However, the salient feature of (6.11) is its simplicity. Since $c_e^{(k)}$ is determined by the network provider, it may also be modified to reflect pricing information. In [12], Lee and Mark have obtained an expression for $c_e^{(k)}$, which takes into account the statistical multiplexing gain, using a linear approximation method.

## Buffer Dimensioning at the Access Node

The effective capacity prescribed by (6.11) is a function of the burst level cell loss rate, $P_e$, and the buffer size, $B$. In an ATM network, the burst level cell loss should be less than $10^{-6}$.[1]

We view the access node as composed of a statistical multiplexer fed by a number of individually leaky bucket controlled sources, as shown is Figure 2. Here, the leaky bucket is structured as having a data buffer and a token pool.

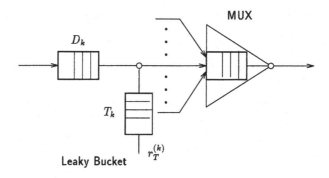

**Figure 2**    Structure of Access Node

The rate of token generation, $r_T^{(k)}$, for the $k$th traffic stream is determined by the effective capacity, $c_e^{(k)}$, i.e., $r_T^{(k)} = c_e^{(k)}$. It is shown in [13] that, as far as the queueing performance of the statistical multiplexer is concerned, it only depends on the total room size of the leaky bucket, i.e., if $D_k$ is the data buffer size and $T_k$ is the token pool size, the performance only depends on the sum $D_k + T_k$, but not on the individual values of $D_k$ and $T_k$. As long as there are tokens in the token pool, data units are consumed. The data buffer starts to accumulate when the token pool is empty, and the leaky bucket departure rate is limited to that of the token generation rate. When the data buffer is full, overflow takes place and newly arriving cells are lost. The rate at which the data buffer overflows is the cell-level loss rate. On the other hand, when the data buffer is empty and there are no new arrivals, the token pool accumulates. When the token pool is full, newly generated tokens are lost. For an *on/off* source, the departure rate is zero during *off* periods, at which times the token

---

[1]The relative importance of the cell level loss and the burst level loss is still a contentious argument within the research community. However, the cell level loss requirement is normally more stringent, e.g., in the neighborhood of $10^{-9}$.

pool accumulates. When the *on* period begins, data cells are being generated a the rate $\gamma_k$, and, since tokens have been accumulated, data cells depart at the rate $\gamma_k$. If the *on* period is longer than the token pool size $T_k$, the departure rate switches to the token generation rate, $r_T^{(k)}$. Otherwise, the departure rate becomes zero, corresponding to an *off* period of the arrival process.

The token generation rate, $r_T^{(k)}$, and the token pool size, $T_k$, represent the service capacity available for the $k$th source stream. The presence of the data buffer mainly provides some smoothing effect on the departure stream, i.e., it behaves as a low-pass filter. Let $X$ be the random variable representing the state of the leaky bucket. Since a data cell is consumed as long as there is a token available in the token pool, the state $X$ then represents the number of data cells in the data buffer, or the number of tokens in the token pool, or 0, which corresponds to the condition that there are neither data cells in the data buffer nor tokens in the token pool. The leaky bucket can be viewed as a finite state machine (FSM) with $(D_k + T_k + 1)$ states. If the arrival process is a renewal type, the stochastic state space developed can be characterized as a Markov chain. Instead of looking into the statistical properties of the departure process in order to analyze the queueing behaviour at the multiplexer, we can focus attention on the number of cells outputted from each of the leaky buckets during a state transition. For the $k$th leaky bucket, it has been formulated in [13] that the leaky bucket behaviour is represented by a state-transition-output matrix, $p^k(z)$, with $(j, i)$th element given by

$$
p_{j,i}^k(z) = \begin{cases} \bar{q}_{D_k-j}^i z^{(-j)^+ +1}, & \text{if } i = D_k \\ q_1^{(k)} z + q_0^{(k)}, & \text{if } j = i = -T_k \\ q_{i-j+1}^{(k)} z^{(-j)^+ - (-i)^+ +1}, & \text{otherwise,} \end{cases} \tag{6.13}
$$

where $q_j^{(k)}$ is the probability of $j$ cells arriving to the $k$th leaky bucket, $\bar{q}_j^{(k)} = 1 - \sum_l^j q_l^{(k)}$, and $(x)^+ = \max(x, 0)$. Let there be $M$ leaky bucket controlled sources. Under a statistical equilibrium condition, the aggregated transition-output matrix of all $M$ leaky bucket controlled sources is given by

$$
P(z) = \otimes_{k=1}^M p^{(k)}(z), \tag{6.14}
$$

where $\otimes$ denotes the Kronecker product [14].

Let $q(l)$ be the probability that the queue length is $l$. Under an infinite buffer assumption and for sufficiently large $l$, the tail of the queue length distribution can be expressed as a linear combination of geometric terms [15]:

$$q(l) = \sum_{\underline{j}=-\underline{T}}^{\underline{D}} \sum_{m=0}^{M-2} \beta_{\underline{j}m} x_{\underline{j}m}^l, \tag{6.15}$$

where $\beta_{\underline{j}m}$ is the coefficient of the spectral expansion and $x_{\underline{j}m}$ is the $(\underline{j}m)$th eigenvalue of the characteristic equation of the probability generating function, $P(z)$. The symbols $\underline{T}$ and $\underline{D}$ represent all the token pool sizes and data buffer sizes. Let $x^*$ be the dominant eigenvalue and $\beta^*$ be the corresponding coefficient of expansion. For sufficiently large $l$, and hence sufficiently small cell loss rate, say in the order of $10^{-9}$, an asymptotic approximation is obtained by retaining only the dominant term in (6.15):

$$q(l) \approx \beta^* x^{*l}. \tag{6.16}$$

Both the dominant root $x^*$ and the corresponding coefficient $\beta^*$ can be determined exactly [15]. However, exact determination of $\beta^*$ requires a knowledge of the boundary values, which can be computationally intensive. It is shown in [15] that the following approximation yields quite accurate results:

$$\beta^* \approx (1-x^*) \frac{1-\rho}{(M-1)/x^* - \varrho'(\tilde{P}(x^*))/\varrho(\tilde{P}(x^*))}, \tag{6.17}$$

where the prime denotes differentiation with respect to $x^*$ and $\varrho(\tilde{P}(x^*))$ denotes the Perron root of the nonnegative matrix $\tilde{P}(x^*)$ [16]. The matrix $\tilde{P}(x)$ is obtained by substituting $z = x^{-1}$ in the aggregated transition-output matrix, $P(z)$, of (6.14). Note that, (6.16), as an approximation of overflow probability, is of the same form as (6.4). By truncating the queue at $l = B$, the cell loss rate as a function of $B$ (CLR(B)) is approximately given by [15]:

$$\text{CLR}(B) = \beta^* \frac{x^{*B}}{1-x^*}. \tag{6.18}$$

## Discussions

For a prescribed burst-level cell loss rate, we can determine the buffer dimension using (6.18). The results can then be used in (6.12) to compute $\zeta$, and hence the equivalent capacity, $c_e^{(k)}$, using (6.11). The schedulable envelope, $\hat{c}^{(i)}$, for the $i$th node is then calculated using (6.2), and the schedulable envelope, $\hat{c}_{\min}$, for admitting a new call is given by the minimum of the schedulable envelope along the network partition, as shown in (6.3). If the effective capacity requested by a new call exceeds the $\hat{c}_{\min}$, the new call is either rejected, or a different effective capacity, and hence different QoS, would be negotiated.

## 4.2 Functionality at the Generic Interior Node

A fundamental difference between the boundary node and an interior node in $\Omega_{np}$ is that the input link speed to an interior node is the same as the output link speed. However, the flow rates, the service priority, and the service capacity allocated to the different traffic streams may differ. The service priority and the resource requirement by the different traffic streams require proper *scheduling* to provide fairness to the various traffic streams. Before attempting to devise a scheduling algorithm, it is necessary to have a good definition of fairness, and to establish a set of criteria that has fairness as a constraint.

As alluded to earlier, traffic entry into the network is limited by the allocated schedulable envelope, which reflects QoS requirement. We expect that the schedulable envelope governs a reference traffic source to exceed the minimum requirement of that traffic source. Thus, the minimum QoS requirement and that offered within the schedulable envelope together provide a range of tolerance, i.e., a range of cell loss rates, cell delays and cell delay variations. Let $\Psi_{np}$ be a set representing all possible ranges of QoS requirements for the network partition. Then, $\psi_k \epsilon \Psi_{np}$ is the tolerance range for the $k$th traffic source.

**Definition 1:** A *service scheduling strategy* is said to be fair if the QoS provided to the $k$th traffic source belongs to the subset $\psi_k$.

The above definition of fairness is reasonable since the individualized schedulable envelope, and hence the QoS subset, $\psi_k$, reflects global network properties.

Our schedulable envelope in terms of the effective capacity, $c_e^{(k)}$, encapsulates the buffer dimension $B$ and the cell loss rate CLR(B). Thus, efficient scheduling is tantamount to having an effective buffer allocation strategy for queue management. Consider a generic node in the context defined earlier. The node handles the service of multi-traffic flows. The complexity of the buffer allocation and queue management strategies increases as the number of traffic flows increases.

### *Elements of the Generic Node*

The generic node is modeled as a single server queueing system with regulated multiple inputs, a buffer of size $B$, and a service capacity $c$. The queue management of the generic node hinges on how the input traffic streams are regulated to offer fair queueing, and on how the total buffer space is partitioned to make efficient use of the available resources. It is proposed to first aggregate the multiple input streams into two classes, in a manner described in [4], and then use an Earliest Due-Date Scheduling scheme to regulate the arrival cells to attain fair queueing. In essence, the operation of the generic node relies heavily on the availability of an effective scheduling scheme and an effective buffer allocation strategy.

## Earliest Due-Date Scheduling

In the present context, an Earliest Due-Date Scheduler is used to regulate each of the two traffic classes, so that arriving cells are queued in the order based on their due date information. To measure the earliness of a cell, a target arrival time is prescribed for each cell as a reference. An Early Arrival Penalty and a dynamic priority adjustment rule are then used to enforce Earliest Due-Date Scheduling to attain fair queueing. The effectiveness of the Earliest Due-Date Scheduling scheme in providing fair queueing is described in [17].

## Buffer Allocation Strategies

From the queue management point of view, it is critically important to represent the multi-traffic flows in a way that service scheduling and buffer allocation is manageable. The scheduling and buffer allocation problem can be made tractable if the multi-traffic flows are aggregated into two types, say using the fluid flow model described in [4]. In the present work, we assume the arrival processes to our generic node have been aggregated into two types, and that each type has been aggregated into a single flow. We therefore have two traffic flows into the buffer of the generic node. The two aggregated flows may have the same service priority or different service priorities. In either case, we can consider a total buffer of size $B$, to be shared by the two traffic flows. We offer below two buffer management strategies.

1. If the two traffic flows have the same service priority, then we still use complete partition, except that the entire buffer of $B$ spaces are completely shared by the two traffic types. When necessary a push-out scheme is used to provide fair allocations. We refer to this method as complete sharing with virtual partition (CSVP).

2. If the two traffic flows have different service priorities, say a low and a high priority, it is proposed to partition the total buffer of size $B$ into distinct

parts, $B_1$ and $B_2$, such that $B = B_1 + B_2$. This method is referred to as complete partition (CP).

## CSVP Scheme for Two Traffic Types of Same Priority (Fluid Flow Model)

The complete sharing with virtual partition scheme to serve two traffic types of the same service priority is depicted in Figure 3. The types 1 and 2 aggre-

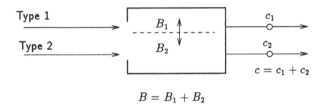

$$B = B_1 + B_2$$

**Figure 3**  Complete Sharing with Virtual Partition Method

gated traffic flows are allocated service capacities $c_1$ and $c_2$, respectively, such that $c = c_1 + c_2$. The system behavior can be characterized by a set of partial differential equations with a triangular boundary.

To characterize the system behavior, it is necessary to first characterize the input processes. A correlated process can be described as having multiple phases. Let $M_1$ and $M_2$ be the number of phases of processes 1 and 2, respectively. Let $\mathbf{H}_1$ and $\mathbf{H}_2$ be the generator matrices of processes 1 and 2, respectively. Also, let $\mathbf{\Lambda} = diag[\lambda_l, \; l = 1, \cdots, M_1]$ and $\mathbf{\Omega} = diag[\omega_k, \; k = 1, \cdots, M_2]$ denote the diagonal rate matrices of processes 1 and 2, respectively. Let $\pi_i$ denote the probability distribution of the phases of process $i$:

$$\pi_i = [\pi_{i1}, \pi_{i2}, ..., \pi_{iM_i}]^t \quad (i = 1, 2).$$

Assume that $\mathbf{H}_1$ and $\mathbf{H}_2$ are irreducible. Then a stationary probability distribution develops, with

$$\mathbf{H}_1^t \pi_1 = \mathbf{0}; \quad \mathbf{H}_2^t \pi_2 = \mathbf{0},$$

where the superscript $t$ denotes matrix transpose.

Let $X_1(t)$ and $X_2(t)$ denote the buffer occupancies of types 1 and 2 at time $t$. Let $s_i(t)$ denote the phase of process $i$, $i = 1, 2$. When $(s_1(t), s_2(t)) = (l, k)$, the traffic are admitted into the buffer at rates $a_1(t)$ and $a_2(t)$, respectively. The admission rates are state-dependent:

1. If $X_1(t) + X_2(t) < B$, $a_1(t) = \lambda_l$ and $a_2(t) = \omega_k$.

2. If $X_1(t) + X_2(t) = B$ and $X_1(t) < B_1$, $a_1(t) = \lambda_l$. In this case, the admission rate for type 2 depends on the relative value of $\lambda_l$ and the service capacity $c$. If $\lambda_l < c$, the type 2 cells are admitted at a rate of $a_2(t) = c - \lambda_l$. If $\lambda_l > c$, the type 2 cells in the buffer are pushed out at a rate of $\lambda_l - c$, i.e., $a_2(t) = c - \lambda_l$ is a negative rate.

3. If $X_1(t) + X_2(t) = B$ and $X_1(t) > B_1$, the admission rates are $a_1(t) = c - \omega_k$, $a_2(t) = \omega_k$.

4. If $X_1(t) + X_2(t) = B$ and $X_1(t) > B_1$, the admission rates are $a_1(t) = c_1$, $a_2(t) = c_2$.

The joint probability distribution functions are defined as

$$F_{lk}(x_1, x_2, t) = P\{s_1(t) = l, \; s_2(t) = k, \; X_1(t) \le x_1, X_2(t) \le x_2\}, \quad x_1 + x_2 < B.$$

If steady state condition is attainable, then

$$F_{lk}(x_1, x_2) = \lim_{t \to \infty} F_{lk}(x_1, x_2, t).$$

In the matrix form

$$\mathbf{F}(x_1, x_2) = [F_{11}(x_1, x_2), \cdots, F_{M-1, M-2}(x_1, x_2)]^t.$$

For the *on/off* processes, the number of phases is 2, and the probability distribution of the phases of process $i$ is

$$\pi_i = [\pi_{i1}, \pi_{i2}]^t, \quad (i = 1, 2).$$

with

$$\mathbf{H}_i^t \pi_i = \mathbf{0}, \quad i = 1, 2.$$

Then, the system is characterized by the following set of partial differential equations [18]:

$$(\mathbf{\Gamma} \otimes \mathbf{I}) \frac{\partial \mathbf{F}}{\partial x_1} + (\mathbf{I} \otimes \mathbf{\Delta}) \frac{\partial \mathbf{F}}{\partial x_2} = (\mathbf{H}_1^t \oplus \mathbf{H}_2^t) \mathbf{F} \quad (x_1 + x_2 < B), \qquad (6.19)$$

where $\mathbf{\Gamma} = \mathbf{\Lambda} - c_1 \mathbf{I}$, $\mathbf{\Delta} = \mathbf{\Omega} - c_2 \mathbf{I}$, $\mathbf{I}$ is an identity matrix, and $\otimes$ and $\oplus$ denote, respectively, the Kronecker product and Kronecker sum of matrices [14]. The

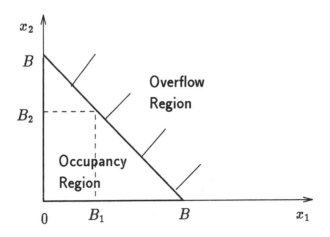

**Figure 4**   Buffer Occupancy for Virtual Partition

buffer occupancy is depicted in Figure 4. The partial differential equation (6.19) is difficult to solve. However, one can obtain an asymptotic solution by decoupling the two traffic types under the assumption of large buffer size, and hence small cell loss probability [18].

Cell loss occurs when the buffer occupancy exceeds the triangular boundary in Figure 4. The detailed analysis can be found in [18]. Here, we make the following observations:

1. When $x_1 = B$ and $x_2 = 0$, only type 1 cells will be lost at a cell loss rate of $(\lambda_l - c_1)$.

2. When $x_1 = 0$ and $x_2 = B$, only type 2 cells will be lost at a cell loss rate of $(\omega_k - c_2)$.

3. When $x_1 + x_2 > B$, both types can lose cells depending on the individual buffer occupancy, as shown in Table 1.

## CP Scheme for Two Priority Classes (Discrete-Time Model)

With the inputs aggregated into a high (class 1) and a low (class 2) priorities, we can choose to partition the total buffer of size $B$ into $B_1$ and $B_2$,

| Type 1 Occupancy | Type 1 | Type 2 |
|:---:|:---:|:---:|
| $0 \leq x_1 < B$ | 0 | $\lambda_l + \omega_k - c_1 - c_2$ |
| $x_1 = B$ | $\lambda_l - c_1$ | $\omega_k - c_2$ |
| $B_1 < x_1 \leq B$ | $\lambda_l + \omega_k - c_1 - c_2$ | 0 |

**Table 1**   Cell Loss Rates when Phases Exceed the Boundary

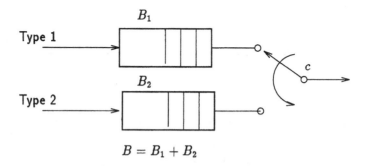

$$B = B_1 + B_2$$

**Figure 5**   Complete Partition for Two Priority Classes

proportional to the percentage of type 1 and type 2 traffic, subject to a higher priority service given to type 1. The buffer partition is shown in Figure 5. Let $\epsilon_1$ denote the fraction of type 1 traffic. Our intuition is that $B_1$ should be a nondecreasing function of $\epsilon_1$ and $B_2$ should be a nondecreasing function of $(1 - \epsilon_1)$. An heuristic partition approach is of the form [19]

$$B_2 = (1 - \epsilon_1)^\mu B \quad \text{and} \quad B_1 = B - B_2,$$

where $\mu$ is the allocation factor. By giving type 1 higher service priority, a value of $\mu$ between 0 and 1 will satisfy the nondecreasing function for $B_1$ and $B_2$.

Let $Q_1(n)$ and $Q_2(n)$ be the number of cells in the type 1 and type 2 queues, respectively, at the end of the $n$th time slot. Let $A_1(n)$ and $A_2(n)$ be the number of type 1 and type 2 cells arriving in the $n$th time slot. We can establish the following recurrence equations to describe the queue evolution:

$$Q_1(n) = \min\{B_1, [Q_1(n-1) - 1 + A_1(n)]^+\}$$

$$Q_2(n) = \begin{cases} \min\{B_2, [Q_2(n-1) - 1 + A_2(n)]^+\} & \text{if } Q_1(n-1) = 0; A_1(n) = 0 \\ \min\{B_2, Q_2 + A_2(n)\} & \text{otherwise} \end{cases}$$

If the arrival processes are renewal, then $(Q_1(n), Q_2(n))$ forms a two-dimensional Markov chain. As mentioned earlier, the *on* and *off* periods of an individual flow in the backbone network is not exponential, even if they were on arrival to the network access point. However, the aggregation of many flows may change this characteristic.

On the assumption that the different traffic flows are independent, the aggregated arrivals may be assumed to obey a binomial distribution, and $(Q_1(n), Q_2(n))$ to form a two-dimensional Markov chain. Then, we can represent the chain by the state transition probability

$$\pi(j, m; i, l) = Pr[Q_1(n) = j, Q_2(n) = m | Q_1(n-1) = i, Q_2(n-1) = l]$$

and steady state probabilities

$$P(j, m) = \lim_{n \to \infty} Pr[Q_1(n) = j, Q_2(n) = m]$$

$$= \sum_{i=0}^{B_1} \sum_{l=0}^{B_2} \pi(j, m; i, l) P(i, l)$$

with the normalization equation

$$\sum_{j=0}^{B_1} \sum_{m=0}^{B_2} P(j, m) = 1.$$

The marginal probabilities are given by

$$P_1(j) = \sum_{m=0}^{B_2} P(j, m); \quad P_2(m) = \sum_{j=0}^{B_1} P(j, m).$$

For given values of $B, \epsilon_1$ and load $\rho$, we can obtain $B_1$ and $B_2$ as a function of $\mu$, and evaluate $P_1(B_1)$ and $P_2(B_2)$, which represent the probabilities of buffer overflow for types 1 and 2, respectively. We can then select the value of $\mu$ to yield the best tradeoff between $P_1(B_1)$ and $P_2(B_2)$. A simple way to estimate $\epsilon_1$ is by taking the relative ratio of the running average of each of the traffic types, i.e.,

$$\hat{\epsilon}_1 = \frac{S_1(n)}{S_1(n) + S_2(n)},$$

where

$$S_1(n) = (1 - \frac{1}{n}) S_1(n-1) + \frac{1}{n} A_1(n)$$

$$S_2(n) = (1 - \frac{1}{n}) S_2(n-1) + \frac{1}{n} A_2(n).$$

## 5  CONCLUSIONS

Traffic management is concerned with usage parameter control (UPC), and traffic scheduling and buffer allocation in the *Transport Plane*. The performance models obtained for the traffic management functions provide a platform for resource allocation and call admission control (CAC) in the *Control Plane*. Our view is that control and scheduling functions can be separately applied, but are coupled to the extent that the enforcement of CAC uses the resource limits imposed by the schedulable envelope. As described in the text, the key to defining the schedulable envelope is through a proper definition of a traffic descriptor, the partitioning of the capacity into schedulable regions, and the determination of an effective capacity that encapsulates the relationship between cell loss probability and the buffer size. Traffic scheduling amounts to devising an effective buffer allocation strategy. We have discussed the aggregation of multi-traffic flows into two types, and presented two buffer allocation strategies to schedule the traffic flows for queueing and service purposes.

## Acknowledgement

This work was supported partially by the Natural Sciences and Engineering Research Council of Canada under Grant No. A7779, a grant from the Canadian Institute for Telecommunications Research (CITR) under the NCE programme of the Government of Canada, and a grant from the Information Technology Research Centre (ITRC) under the Centres of Excellence programme of the Province of Ontario.

## REFERENCES

[1] The ATM Forum, "ATM User-Network Interface Specification," Version 3.0, 1993.

[2] Turner, J. S., "New Directions in Communications (or Which Way to the Information Age?)," IEEE Commun. Mag., Vol. 24, No. 10, pp. 8-15, Oct. 1986.

[3] Kosten, L., "Stochastic Theory of Data Handling Systems with Groups of Multiple Sources," in *Performance of Computer-Communication Systems*, eds. H. Rudin and W. Bux, Elsevier, 1984, pp. 321-331.

[4] Lee, H. W. and J.W. Mark, "ATM Network Traffic Characterization Using Two Types of *on/off* Sources," Proc. of IEEE INFOCOM '93, pp. 152-159, March 28-Apr.1, 1993, San Francisco, CA.

[5] Ren, J. -F., J. W. Mark and J. W. Wong, "End-to-End Performance in ATM Networks," IEEE ICC '94 Conference Record, May 1-5, 1994, New Orleans, La, pp. 996-1002.

[6] Anick, D., D. Mitra and M. M. Sondhi, "Stochastic Theory of a Data-Handling System with Multiple Sources," The Bell System Tech. J., Vol. 61, No. 8, pp. 1871-1894, Oct. 1982.

[7] Gibbens, R. J. and P. J. Hunt, "Effective Bandwidths for the Multi-type USA Channel," Queueing Systems, Vol. 9, pp. 17-27, 1991.

[8] Kesidis, G., J. Walrand and C. -S. Chang, "Effective Bandwidths for Multiclass Markov Fluids and Other ATM Sources," IEEE/ACM Trans. on Networking, Vol. 1, No. 4, pp. 424-428, Aug. 1993.

[9] Stern, T. E. and A. I. Elwalid, "Analysis of Separable Markov-Modulated Rate Models for Information-Handling Systems," Adv. Appl. Prob., No. 23, pp. 105-139, 1991.

[10] Elwalid, A. I., D. Mitra and T. E. Stern, "Statistical Multiplexing of Markov Modulated Sources: Theory and Computational Algorithms," Proc. ITC-13, pp. 495-500, Copenhagen, 1991.

[11] Elwalid, A. I. and D. Mitra, "Effective Bandwidth of General Markovian Traffic Sources and Admission Control of High-Speed Networks," IEEE/ACM Trans. on Networking, Vol. 1, No. 3, pp. 329-343, June 1993.

[12] Lee, H. W. and J. W. Mark, "Capacity Allocation in Statistical Multiplexing of Heterogeneous Sources," Proc. of IEEE INFOCOM '94, June 1994, Toronto, Canada, pp. 708-715.

[13] Ren, J. -F., J.W. Mark and J.W. Wong, "Performance Analysis of a Leaky-Bucket Controlled ATM Multiplexer," Performance Evaluation, special issue on *Bandwidth Management and Congestion Control of High-speed Networks*, Vol. 19, No. 1, pp. 73-101, Jan. 1994.

[14] Graham, A., *Kronecker Product and Matrix Calculus: with Applications*, John Wiley & Sons, 1981.

[15] Ren, J. -F., J. W. Mark and J. W. Wong, "Asymptotic Analysis of a Leaky Bucket Controlled ATM Multiplexer," IEEE ICC '93 Conference Record, May 23-26, 1993, Geneva, Switzerland, pp. 1386-1390.

[16] Henrici, H., *Applied and Computational Complex Analysis*, Vol. 1, John Wiley and Sons, New York, 1974.

[17] Ren, J. -F., J. W. Mark and J. W. Wong, "A Dynamic Priority Queueing Approach to Traffic Regulation and Scheduling in B-ISDN," to appear in IEEE GLOBECOM '94 Conference Record, Nov. 27-Dec. 1, 1994, San Francisco, CA.

[18] Wu, G.-L. and J.W. Mark, "A Buffer Allocation Scheme for ATM Networks: Complete Sharing Based on Virtual Partition," Proc. of ITC-sponsored Seminar on Teletraffic Analysis for Current and Future Telecom Networks, Nov. 15-19, 1993, Bangalore, India, pp. 19-26.

[19] D.X. Chen and J.W. Mark, "Delay and Loss Control of Output-Buffered Fast Packet Switches Supporting Integrated Services," IEEE SUPERCOMM/ICC '92 Conf. Record. vol.2, pp. 985-989, June 14-18, 1992, Chicago, IL.

# 7

# DISCRETE-TIME ANALYSIS OF USAGE PARAMETER CONTROL FUNCTIONS IN ATM SYSTEMS

## P. Tran-Gia

*Institute of Computer Science, University of Würzburg,*
*Am Hubland, D-97074 Würzburg, Germany*

## ABSTRACT

The design of the User Network Interface (UNI) in accordance with the incorporated Usage Parameter Control (UPC) plays an important role in the current ATM development and standardization process. Due to the discrete-time nature of ATM cell traffic and the control functions at UNI, queueing models operating in discrete-time domain can be used in a quite direct way.

In this paper, two models dealing with UNI/UPC functions will be presented i) a queueing model to analyse the cell traffic shaping using a spacer, where a discrete-time algorithm for the spacer output process is developed and ii) a queueing model for the *generic cell rate algorithm* (GCRA) is derived, with which dimensioning aspects of the *cell delay variation* (CDV) are discussed, in accordance with a versatile discrete-time algorithm.

# 1 DISCRETE-TIME MODELLING AND ANALYSIS

## 1.1 General

Discrete-time models appear more and more frequently in performance evaluation of modern communication systems and play an increasingly important role. On the one hand, new system structures and principles often employ discrete or discretized basic time and data units. As discussed above, examples are the concept of cells in ATM networks or time slots in high-speed local and

111

metropolitan area networks (e.g., DQDB). On the other hand, system parameters and input values are often based on measured data, which are given in the form of histograms. They are discrete-time by nature. These facts lead to the development of discrete-time models in performance analyses in the recent literature.

For the analysis of this class of models, conventional methods operating in continuous time are obviously inappropriate. Due to the lack of discrete-time methods they are used in some cases in an approximate sense. In these studies equivalent continuous-time model components are employed, e.g. the discrete-time stochastic arrival and service processes are approximately described by means of random variables with well-known time-continuous types of distribution functions. A small number of studies [1, 18, 20, 21, 27, 28, 29, 31] deals with direct analysis approaches for discrete-time models. Most of these studies take into account the discrete-time analysis of basic queueing models like single server systems [1, 20, 21, 27] or queueing networks. Some other studies present discrete-time analysis of general polling systems, overload control models in communication switching systems [28], routing mechanisms [31] or multiplexing schemes in modern communication system architectures [29]. A comprehensive survey can be found in [18].

The main purpose of this paper is to illustrate the use of discrete-time analysis techniques, considering models dealing with usage parameter control (UPC) functions in ATM systems at the user-network interface (UNI). In these analyses, performance measures of interest are usually the cell blocking probability (e.g. in the range of $10^{-5}$ to $10^{-9}$) or the waiting time distribution function with the percentiles. Thus, simulation methods normally reach their limits very fast due to excessive computing time, and the need of appropriate analysis techniques is obvious.

This paper is organized as follows. In the subsection to follow basic discrete-time analysis techniques are outlined, where the GI/GI/1 queue with bounded delay is taken as example. This model is the underlying model for the performance evaluation of the submodels in the next section. Section 2 presents performance analyses of UPC functions in ATM systems, where two submodels are discussed. The first submodel is used to investigate the traffic shaping function using a generic cell process spacer, where the characterization of cell traffic passing the spacer is taken into account. The second submodel deals with the dimensioning of cell delay variation tolerance, which appears in the generic cell rate algorithm used in the user-network interface of ATM networks.

## 1.2 The GI/GI/1 queue with bounded delay

We consider in this subsection the case of the GI/GI/1 system with bounded delay, i.e. the waiting time of customers is limited to a maximum value of $L$. Customers who arrive and would have to wait longer than a threshold value, say $L-1$, are rejected. The term discrete-time indicates here that the time axis is slotted in equidistant time units of length $\Delta t$. The service time $B$ is generally distributed and the arrival process is a general stochastic process characterized by a generally distributed interarrival time $A$. Waiting customers in the queue will be treated according to a first-in, first-out (FIFO) service discipline.

The main performance measures of interest are here the distribution functions of the waiting time and the inter-departure interval. We refer to studies appearing in the literature dealing with the calculation of the waiting time distribution function of the GI/GI/1 queue [1, 19, 20, 21, 22, 26, 27]. Most of these methods are related to solutions of the Lindley integral equation [22], which is a special form of Wiener-Hopf equations.

This modification of the basic GI/GI/1 system has been used in modelling of overload control strategies in switching systems [29] and in backpressure mechanisms in reservation-based access mechanisms [30]. In this paper, this model will be used for performance evaluation of UPC functions in ATM systems like the spacing device and the peak cell-rate monitoring algorithm.

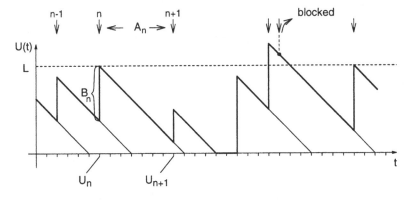

**Figure 1** Sample path of GI/GI/1 queue with bounded delay

As mentioned above, the random variables (RVs) are of discrete-time nature, i.e. the time axis is divided into intervals of unit length $\Delta t$. As a consequence, samples of those random variables are integer multiples of $\Delta t$; the time discretization is equidistant. We further assume that the discrete distributions have finite lengths. For a discrete-time random variable $X$, we use the notation $x(k) = \Pr\{X = k\}, -\infty < k < +\infty$, for the distribution (probability mass function) of $X$, $X(k) = \sum_{i=-\infty}^{k} x(i), -\infty < k < +\infty$, for the distribution function of $X$, $EX$ for the mean and $c_X$ for the coefficient of variation of $X$.

We denote by $A_n$ the RV for the interarrival time between the $n$-th and the $(n+1)$-st customer, $B_n$ the RV for the service time of the $n$-th customer and again, $U_n$ the RV for the unfinished work in the system immediately prior to the $n$-th customer arrival.

A snapshot of the state process development in the system is shown in Fig. 1. Observing the $n$-th customer in the system and the condition for customer acceptance upon arrival instant, the following *conditional random variables* for the workload seen by an arriving customer are introduced:

$$U_{n,0} = U_n | U_n < L \qquad\qquad U_{n,1} = U_n | U_n \geq L, \qquad (7.1a)$$

$$U_{n+1,0} = U_{n+1} | U_n < L \qquad U_{n+1,1} = U_{n+1} | U_n \geq L. \qquad (7.1b)$$

The distributions of these random variables, adjusted by normalization, are:

$$u_{n,0}(k) = \frac{\sigma^{L-1}[u_n(k)]}{\Pr\{U_n < L\}} = \frac{\sigma^{L-1}[u_n(k)]}{\sum_{i=0}^{L-1} u_n(i)} \qquad (7.2a)$$

$$u_{n,1}(k) = \frac{\sigma_L[u_n(k)]}{\Pr\{U_n \geq L\}} = \frac{\sigma_L[u_n(k)]}{\sum_{i=L}^{\infty} u_n(i)} \qquad (7.2b)$$

where $\sigma^m(.)$ and $\sigma_m(.)$ are operators which truncate parts of a probability distribution function. The results of these operations are unnormalized distributions defined by:

$$\sigma^m[x(k)] = \begin{cases} x(k) & k \leq m \\ 0 & k > m \end{cases} \qquad (7.3a)$$

$$\sigma_m[x(k)] = \begin{cases} 0 & k < m \\ x(k) & k \geq m \end{cases} \tag{7.3b}$$

Observing the development of the process (cf. Fig. 1) together with the maximum delay $L-1$, the following relationships between the RVs and their distributions are obtained:

1. $U_n < L$: customer acceptance

$$U_{n+1,0} = U_{n,0} + B_n - A_n \tag{7.4a}$$
$$u_{n+1,0}(k) = \pi_0[u_{n,0}(k) \star b_n(k) \star a_n(-k)]. \tag{7.4b}$$

2. $U_n \geq L$: customer rejection

$$U_{n+1,1} = U_{n,1} - A_n \tag{7.5a}$$
$$u_{n+1,1}(k) = \pi_0[u_{n,1}(k) \star a_n(-k)]. \tag{7.5b}$$

In these equations, the symbol "$\star$" denotes the discrete convolution operation:

$$a_3(k) = a_1(k) \star a_2(k) = \sum_{j=-\infty}^{+\infty} a_1(k-j) \cdot a_2(j) \tag{7.6}$$

and $\pi_0(.)$ the following operator:

$$\pi_0(x(k)) = \begin{cases} x(k) & k > 0 \\ \sum_{i=-\infty}^{0} x(i) & k = 0. \\ 0 & k < 0 \end{cases} \tag{7.7}$$

The distribution of the workload seen by the $(n+1)$-st customer is:

$$u_{n+1}(k) = \Pr\{U_n < L\} \cdot u_{n+1,0}(k) + \Pr\{U_n \geq L\} \cdot u_{n+1,1}(k). \tag{7.8}$$

From eqns. (7.4b), (7.5b) and (7.8), we finally arrive at a recursive relation to calculate the workload at arrival epochs of customers:

$$
\begin{aligned}
u_{n+1}(k) &= \pi_0[\sigma^{L-1}[u_n(k)] \star b_n(k) \star a_n(-k)] + \pi_0[\sigma_L[u_n(k)] \star a_n(-k)] \\
&= \pi_0[(\sigma^{L-1}[u_n(k)] \star b_n(k) + \sigma_L[u_n(k)]) \star a_n(-k)]
\end{aligned}
\tag{7.9}
$$

Using this equation an algorithm to calculate the the workload prior to customer arrivals can be found. The algorithm can be used for both stationary and nonstationary traffic conditions. Under stationary conditions the index $n$ and $(n + 1)$ in this equation can be suppressed. The computational diagram is depicted in Fig. 2.

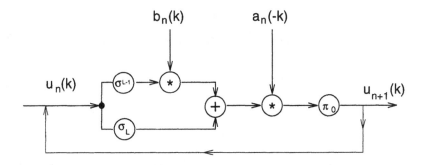

**Figure 2**   Computational diagram for GI/GI/1 queue with bounded delay

Finally, the customer rejection probability in statistical equilibrium is:

$$
B = \sum_{i=L}^{\infty} u(i).
\tag{7.10}
$$

This performance measure will be used e.g. to compute the probability for a cell to be conforming due to the generic cell rate algorithm as discussed in the next section.

# 2 ANALYSIS OF USAGE PARAMETER CONTROL FUNCTIONS

## 2.1 User-network interface and usage parameter control

In the current standardization process of ATM networks the design issues concerning the user-network interface (UNI) in accordance with the usage parameter control (UPC) are crucial points. The main functions of the UPC/UNI illustrated in Fig. 3 are under discussion in standardization bodies (cf. CCITT recommendation [7]). We will briefly describe major UNI/UPC functions in the following with focus to modelling aspects.

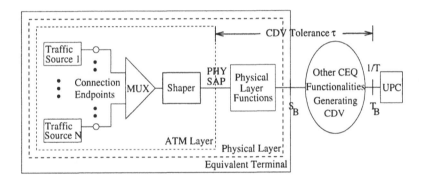

**Figure 3** Reference configuration from CCITT Draft Rec. I.371

Fig. 3 shows a number of virtual channel connections (VCC) multiplexed at the network edge. The multiplexed cell stream has to pass physical layer functions and is multiplexed further with operation and maintenance (OAM) cell traffic before entering the ATM network. Due to the connection admission control (CAC) a VCC is accepted or rejected. In the case of being accepted by the network, the VCC is thought of to have a contract with the network: the VCC user agrees to keep the negotiated traffic characteristics during the connection holding time, the network guarantees a predefined quality of service (QoS). Once a connection is established, policing is provided to guarantee the desired QoS for all connections according to their traffic contracts.

In the first phase of ATM system implementation and testing the most important connection parameter is the *peak cell rate*; this parameter is already standardized by various standardization bodies. The peak cell rate $R_i$ of the VCC $i$ is defined as the inverse of the minimum time $T_i$ between the emissions of two cells from this connection. Currently there are discussions about the introduction of a *sustainable cell rate* in the standards. A preliminary definition of this measure can be found e.g. in Draft Rec. I.371 of CCITT ([6, 7]).

We observe now the cell stream of a connection crossing the multiplexer to enter the ATM network, as shown in Fig. 4. The traffic process I generated in higher protocol layers has to pass a spacing device, which ensures that the peak cell rate $R_i$ of the VCC $i$ is not exceeded. This results in the cell process marked by II in Fig. 4. In [5, 27, 30] a *cell spacer* was suggested as traffic shaper. Another approach using a *spacing policer* was presented in [7]. Different virtual channel connections with different peak cell rates may be multiplexed. It should be noted that the arising spacer delay can affect the end-to-end performance drastically. This phenomenon will be the major subject of the performance model in section 2.2.

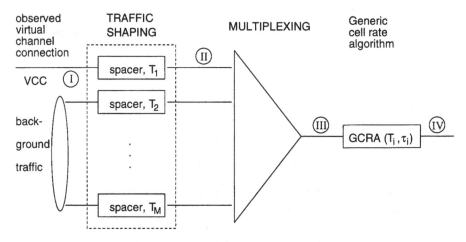

**Figure 4**    Cell traffic spacing and cell delay variation

According to the multiplexing function and to additional superposition of OAM cell traffic, a cell stream, which has been already spaced, can be again perturbed. At point III, the minimum distance $T_i^\star$ between cells which belong to a connection $i$ can be smaller than $T_i$ and thus the peak cell rate $R_i^\star$ at point III can momentarily be higher than $R_i$. This effect is often refered to as the *cell delay variation* (CDV), which can be observed in cell process modifying phenomena like cell clumping or cell dispersion, which strongly influence the cell process characteristics and the short-term rate of a VCC through the ATM network and thus, its quality of service provided by the ATM system [3]. Cell clumping can lead to short-term network congestions. Cell dispersion causes the necessity for larger buffer at the receiving site of the ATM network if the customer equipment has tight delay constraints (e.g. play-out buffer for voice/video). In a more general context, CDV can be introduced by different factors, e.g. i) multiplexing cells from different ATM connections, ii) UPC and *network parameter control* (NPC), iii) segmentation and reassembly functions in the ATM adaptation layer (AAL) and iv) other network and protocol functionalities.

To control the cell delay variation, the generic cell rate algorithm (GCRA) is introduced. By applying the GCRA to control the peak cell rate, it is also called the *virtual scheduling algorithm* (VSA). It can be shown that this algorithm is equivalent to a *continuous-state leaky bucket algorithm* (cf. [6]). The main function of the GCRA is to limit the CDV by controlling the cell stream and to declare accordingly cells during a heavy-traffic phase to be *conforming* or *non-conforming*. This is done connection-wise using two parameters: the minimal inter-cell interval $T_i$ and the CDV tolerance $\tau_i$. Thus, the algorithm is denoted by GCRA($T_i,\tau_i$). Details of the algorithm will be described in Section 2.3, where an analysis to calculate the probability for a cell to be non-conforming is presented. The analysis also delivers the distribution function of the inter-cell process at point IV (cf. Fig. 4).

It should be noted that the location and the order of the three basic function blocks i) the spacer, ii) the multiplexer and iii) the GCRA as shown in Fig. 4 represent only one of various architectural possibilities, which are organized in accordance with the ATM switching system design. The GCRA can be performed e.g. at the same time with the multiplexing function.

Details about the operation of the cell spacer and the generic cell rate algorithm will be discussed later in this section.

## 2.2   Analysis of a cell spacer

The aim of the performance study in this section is an analytical treatment
of a spacing device used for cell shaping functions. The cell process spacer
is often discussed in recent literature [4, 5, 11, 13, 32]. The analysis is based
on a discrete-time GI/D/1 queueing system with bounded delay. The cell
arrival process which is subject to spacing can be chosen arbitrarily. The algorithm aims at the calculation of the spacer output process in terms of the
cell inter-departure time distribution, which gives insights to understand the
traffic stream forming properties of the spacing mechanism. It should be noted
that the spacer output process is in general non-renewal; its description using
distributions will be done in a cumulative sense.

### *Traffic shaping using a spacer*

As discussed above, the peak cell rate $R_i$ of a virtual channel connection in
ATM is defined as the inverse of the minimum cell inter-arrival time $T_i$. This
parameter is part of the traffic contract between the VCC and the network.
The basic function of a cell spacer is to keep the cells generated by a VCC to
be at least $T_i$ apart. This is illustrated in Fig. 5.

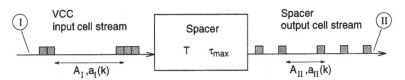

**Figure 5**   Basic spacer model

The time axis is discretized by the cell duration $\Delta t$. The cell process of the VCC
is described by the arbitrarily distributed random variable $A_I$ (cf. the mark I
in Fig. 4 and Fig. 5) for the cell interarrival time. The spacer output process
is represented by the discrete-time RV $A_{II}$ for the inter-departure time. The
minimum inter-cell interval for the observed VCC is $T$. We consider further a
maximum spacer delay $\tau_{max}$; cells which would have to wait in the spacer for
longer than $\tau_{max}$ are rejected.

### *Spacer modelling with bounded delay GI/D/1*

We introduce the random variable $U(t)$ for the time-dependent spacer state,
which stands for the amount of unfinished work in the spacer at time $t$. Accord-

ing to the spacing scheme a cell which arrives at time $t_0$ and sees the spacer in the state $U(t_0) = k$ has to wait $k$ time units in the spacer before being issued.

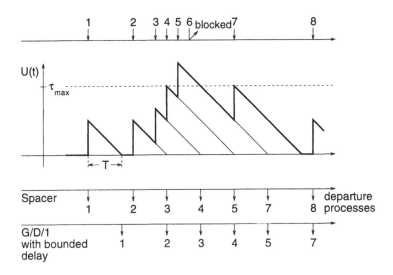

**Figure 6** Spacer state process and output stream

A sample path of $U(t)$ is depicted in Fig. 6, where the spacer input process, the spacer output process and — for comparison purposes — the output process of a GI/D/1 queue with bounded delay are shown. Cells 1 and 2 find the spacer empty and pass it without delay. Cells 3, 4 and 5 arrive at the spacer and see a positive spacer state $U(t)$; they have to be spaced according to the value of $U(t)$ upon their arrivals. Cell number 6 would have to wait longer than $\tau_{max}$ and is thus rejected, since it is a non-conforming cell. It can be seen that an accepted cell increases $U(t)$ by an amount of $T$, which is thought of as the *virtual service time* of the spacing process.

As can be observed in Fig. 6, the output process of the spacer is *similar* to the output of an equivalent GI/D/1 queue, where a delay of $T$ is the only difference. This fact is also used in [32]. A spacer analysis using a $GI^{[X]}/D/1$ queue, which is similar to the discrete-time analysis described here, can be found in Hübner et al. [15].

The analysis used to deliver the results in this section is summarized as follows. For the case of a spacer with unlimited delay ($\tau_{max} \to \infty$) the equivalent model is a GI/D/1 queue, for which the analysis in both time and transform domains

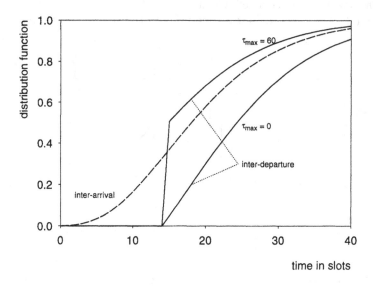

**Figure 7**   Illustration of cell spacing function

as discussed in the first section, can be used. Given the case of a spacer with bounded delay, the state analysis as presented in the previous section can be employed. It should be noted that for the calculation of the output process the use of the equivalent queue of type $GI^{[X]}/D/1$ is more advantageous.

## Dimensioning issues

To illustrate the influence of the spacer on the cell process we consider a spacer, which enforces the peak cell rate of a VCC with a minimum inter-cell interval of $T = 15$. The input process $A_I$ is assumed to be negative-binomially (cf. [27]) distributed.

Fig. 7 depicts how the spacer affects and forms the cell process for the case of $E[A_I] = 20$ and $c_{A_I} = 0.5$, where the distribution functions of the input and the output process of the spacer are shown. This is done for different values of maximum spacer delay $\tau_{max}$. The shape of the inter-departure distribution function indicates the truncation function performed by the spacer.

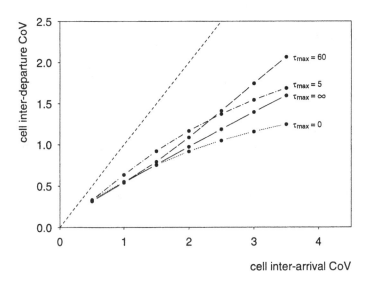

**Figure 8**   Process smoothing by cell spacing

Fig. 8 illustrates the main function of the spacer, i.e. to smooth a bursty, high-variance input cell stream. To illustrate the smoothing effect quantitatively, we take the case of $E[A_I] = 6$ and draw the coefficient of variation $c_{A_{II}}$ of the inter-departure process as a function of the coefficient of variation of the input cell stream. As expected, $c_{A_{II}}$ is smaller than $c_{A_I}$ due to the cell spacing function. This effect is observed for all values of the maximum spacer delay.

## 2.3   Cell Delay Variation modelling

The aim of the performance study in this section is a discrete-time analysis of the generic cell rate algorithm, which is used to monitor the peak cell rate of a VCC in ATM environments. The analysis is again based on a discrete-time GI/D/1 queueing system with bounded delay. The cell arrival process to be controlled by the GCRA can be arbitrarily chosen. The purpose is the calculation of the probability $B$ for a cell to be non-conforming. Subsequently, dimensioning aspects of the cell delay variation tolerance in the UPC are studied.

## The Generic Cell Rate Algorithm

As discussed above, CDV can be introduced by various factors like the multiplexing function, the UPC, the NPC or the segmentation and reassembly functions. To prevent the cell clumping and dispersion to cause network congestion, the GCRA is recommended [6], which has been first applied to monitor the peak cell rate.

The GCRA, as depicted in Fig. 4, monitors each VCC individually. We take into account the virtual channel connection $i$ and suppress the index $i$ in the notation $GCRA(T_i, \tau_i)$. Thus the parameters of the $GCRA(T, \tau)$ for the VCC $i$ are the target minimum inter-cell interval $T$ and the CDV tolerance $\tau$. The main problem is to scale $\tau$ for a VCC in such a way that the probability for a cell to be non-conforming (and rejected or tagged) is low while the CDV is reduced. At the same time, $\tau$ must be dimensioned to prevent overload caused by non-conforming VCC's.

The working mode of the $GCRA(T, \tau)$ is as follows. The algorithm determines whether a cell is generated too close to the last cell, indicating that the connection generates cells with a higher rate than the negotiated rate, or not. Distinction is made between the *theoretical arrival time* $TAT_j$ and the *actual arrival time* $t_j$ of a cell $j$. According to the characteristic of the cell process of the VCC $i$ (marked by III in Fig. 4), cells can be marked as *conforming* or *non-conforming*. The way to treat the next cell (number $(j + 1)$) is as follows:

1. Estimate the theoretical arrival time $TAT_{j+1}$ of cell $j + 1$

   (a) $TAT_{j+1} = t_j + T$   if   $t_j \geq TAT_j$:
       cell $j$ is generated after its expected arrival. It should be noted that the late generation of this cell does not allow for an earlier generation of the next cell.

   (b) $TAT_{j+1} = TAT_j + T$   if   $t_j < TAT_j$ and cell $j$ is conforming:
       cell $j$ is generated prior to its expected arrival but is still within the CDV tolerance. The TAT of the next cell is set as if cell $j$ had been generated at its TAT and not earlier.

   (c) $TAT_{j+1} = TAT_j$   if   $t_j < TAT_j$ and cell $j$ is non-conforming:
       cell $j$ is generated prior to its expected arrival and lies outside the tolerance. Cells which are identified as non-conforming can optionally be tagged or rejected (cf. [6]). It is assumed here that non-conforming cells are discarded. The TAT for the next cell is not modified in this case.

2. The next cell $j + 1$ arrives at $t_{j+1}$. It will be considered as

   (a) conforming if $t_{j+1} \geq TAT_{j+1} - \tau$

   (b) non-conforming if $t_{j+1} < TAT_{j+1} - \tau$.

The algorithm guarantees that cells from a VCC enter the ATM network at the $T_B$ reference point (cf. Fig. 3) with a long term rate of at most $1/T$. It should be noted that the tolerance $\tau$ could be chosen to be larger than the target minimum inter-cell time $T$. Furthermore, due to the CDV tolerance $\tau$, the smallest inter-cell interval of the cell process at point IV in Fig. 4 is tolerated by the GCRA to be shorter than the (theoretical) target minimum inter-cell interval $T$. The peak cell rate $1/T$ could momentarily be exceeded and deliberately violated.

## Model of the Generic Cell Rate Algorithm

The discrete-time random variable $U(t)$ for the time-dependent GCRA state is introduced, which represents the remaining time to the next theoretical arrival time. The value of $U(t)$ can be thought of as the *virtual unfinished work* according to the GCRA function. A cell which arrives at time $t_0$ and sees the GCRA in the state $U(t_0) = k$ will be considered as non-conforming for $k \geq \tau$ and conforming for $k < \tau$.

A sample path of $U(t)$ is depicted in Fig. 9, where the theoretical arrival time $TAT_j$, the actual arrival time $t_j$ and the tolerated early arrival time (control measure) $TAT_j - \tau$ are shown. Cells 1 and 2 arrive after their TAT and are conforming. The actual arrival time of cell 3 lies before its TAT but is still within the tolerance (i.e. after $TAT_3 - \tau$); cell 3 is therefore conforming. Cell 4 arrives at the GCRA before the tolerance interval ($t_4 < TAT_4 - \tau$) and is non-conforming.

As can be observed in Fig. 9, the state process of the virtual unfinished work of the GCRA is identical to the unfinished work process of a GI/D/1 queue with bounded delay. A conforming (and accepted) cell increases $U(t)$ by an amount of $T$.

The modelling approach here was first presented by Hübner [14]. The assumptions made are quite general. The cell process of the VCC entering the GCRA function is described by the arbitrarily distributed random variable $A_{III}$ (cf. the mark III in Fig. 4) for the cell interarrival time. The GCRA output process

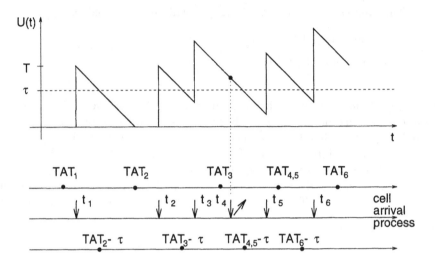

**Figure 9** Generic Cell Rate Algorithm state process

is represented by the discrete-time RV $A_{IV}$ for the interdeparture time. In general, this process is non-renewal. Its distribution function given in this analysis is a cumulative one. To compute the probability for a cell to be non-conforming the blocking probability given in eqn. (18) can be taken.

## Dimensioning issues of Generic Cell Rate Algorithm

In the following we take results from [14] to illustrate the use of the analysis. The traffic scenario depicted in Fig. 4 is considered, where the GCRA is used to monitor the cell process generated by a constant bit rate (CBR) source. The background traffic is characterized by a negative-binomially distributed inter-cell interval $X$. A hybrid approach using both simulation and analysis is employed in [14], where the inter-cell distribution of the process $A_{III}$ for the CBR source after being multiplexed with the background traffic is obtained using simulation.

Unless noticed otherwise, CBR source traffic is assumed to have higher priority by the multiplexer as background traffic. In the numerical results the inter-cell time of the observed CBR source is chosen at $T^\star = 24$. If the CBR source fulfills the traffic contract, its cell rejection probability $B$ should be below a defined quality of service, say $10^{-9}$, when the monitor parameters $T$ and $\tau$ are chosen appropriately. For this range of blocking probabilities, analytical methods are advantageous. Normally, without an elastic tolerance, the parameter $T$ in the $GCRA(T, \tau)$ is the same as $T^*$. In [4] is was stated that in some cases $T$ must be chosen significantly smaller than $T^*$ to be able to meet the performance requirements. In Table 1, taken from [14], minimum values of $\tau$ are listed for different choices of $T$ to keep $B$ below $10^{-9}$.

**Table 1** Dependency of minimum $\tau$ on $T$ (for $B < 10^{-9}$)

| $T =$ | 23 | 22 | 21 | 20 |
|---|---|---|---|---|
| $E[X] = 4, c_X = 1, \tau \geq$ | 8 | 5 | 3 | 1 |
| $E[X] = 4, c_X = 2, \tau \geq$ | 37 | 29 | 26 | 23 |
| $E[X] = 2, c_X = 1, \tau \geq$ | 18 | 14 | 11 | 9 |
| $E[X] = 2, c_X = 2, \tau \geq$ | 107 | 65 | 50 | 42 |

It can be clearly seen that $\tau$ depends not only on the background traffic intensity, as already observed in [13], but is also strongly influenced by the coefficient of variation $c_X$. This has to be incorporated in the dimensioning of the CDV tolerance as illustrated in Fig. 10, where the influence of varying $\tau$ and $c_X$ on the probability $B$ for CBR cells to be non-conforming, with $T^* = 24$, $E[X] = 2$, and $T = 20$, is shown.

It can be observed in Fig. 10 that the cell blocking probability can be remarkably reduced by increasing the tolerance value $\tau$. However, a choice of larger $\tau$ would also enlarge the burst length with higher cell rate than the target peak cell rate in the VCC contract.

If the amount of non-conforming cells is too large although the traffic contract is fulfilled by the VCC user (e.g. due to high CDV caused by multiplexing stages), there are different possibilities to scale the parameters of the GCRA function to monitor the agreed cell blocking probability $B$. The first possibility is to dimension $T < T^*$, which means the introduction of an additional elastic tolerance. This does not seem to be able to be incorporated in the standards. The second possibility is to increase the CDV tolerance $\tau$. This must be done in a careful way. If $\tau$ is too large, the number of back-to-back cells which are still recognized as conforming cells increases. As a consequence, the cell clumping effect arises, which can lead to network congestion. Furthermore, for a large $\tau$, cells from a traffic source which violates the traffic contract can not be appropriately controlled.

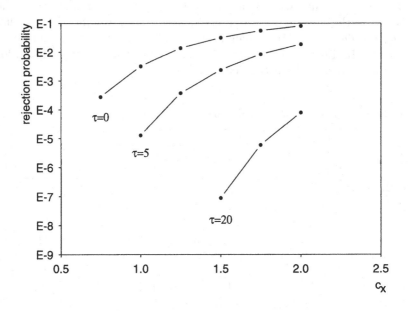

**Figure 10**   Dimensioning of CDV tolerance

# 3   CONCLUSION AND OUTLOOK

The purpose of this paper is to present methods of discrete-time modelling and analysis, which gain importance due to the development of modern computer networks and communication systems operating with fixed-size packets like cells in ATM systems or slots in DQDB subnetworks.

The analysis technique and related algorithms have been presented by means of the treatment of the discrete-time GI/GI/1 queue and GI/GI/1 queue with bounded delay, where algorithms dealing directly with probability distributions in time domain and analysis techniques in transform domain have been outlined.

Subsequently, discrete-time performance analyses of usage parameter control functions in ATM systems have been presented. Submodels for the traffic shaping function using a cell process spacer and for the dimensioning of cell delay variation tolerance in conjunction with the generic cell rate algorithm used in the user-network interface of ATM networks have been discussed.

It has been shown that discrete-time models are well-suited for this class of models. In particular, the discrete-time analysis technique delivers not only mean values but also the entire distribution functions of the cell traffic processes. Furthermore, in modelling cases where very small blocking probabilities are subjects of the performance investigation, discrete-time analysis is advantageous compared to simulations.

## Acknowledgements

The author would like to thank Frank Hübner for stimulating discussions and for the numerical results. Helps and careful reviews provided by Sandra Heilmann, Michael Ritter, Alexander Schömig and Kurt Tutschku are appreciated.

# REFERENCES

[1] Ackroyd M.H., "Computing the waiting time distribution for the G/G/1 queue by signal processing methods", IEEE COM-28(1980) 52-58.

[2] ATM Forum, "Traffic Control and Congestion Control", Draft Baseline Document, April 1993.

[3] Bernabei F., Gratta L., Listanti M., Sarghini A., "Analysis of ON-OFF Source Shaping for ATM Multiplexing", INFOCOM 1993, pp. 11a.3.

[4] Boyer P., Guillemin F.M., Servel M.J., Coudreuse J.-P., "Spacing Cells Protects and Enhances Utilization of ATM Network Links", IEEE Network Vol. 6(1992) No.5.

[5] Brochin F.M., "A Cell Spacing Device for Congestion Control in ATM Networks", Performance Evaluation, Vol.16, No.1-3, November 1992, 107-127.

[6] CCITT Draft Recommendation I.371, "Traffic Control and Congestion Control in B-ISDN", June 1992.

[7] CCITT Study Group XVIII Contribution D.2373, "A Proposal for a Definition of a Sustainable Cell Rate Traffic Descriptor", January 1993.

[8] COST 224 Final Report, "Performance evaluation and design of multiservice networks", J.W. Roberts (editor), Commission of the European Communities (EUR 14152), October 1991.

[9] Daley D.J., "Notes on Queueing Output Processes", Math. Methods in Queueing Theory, Springer, 1974, 351-358.

[10] Gravey A., Boyer P., "Cell Delay Variation Specification in ATM Networks", Proc. IFIP Workshop TC6, Modelling and Performance Evaluation of ATM Technology, La Martinique, January 1993.

[11] Guillemin F.M., Boyer P., Romoeuf L., "The Spacer-Controller: Architecture and First Assessments", Workshop on Broadband Communications, Estoril, Portugal, January 1992, 313-323.

[12] Guillemin F.M., Monin W., "Limitation of Cell Delay Variation in ATM Networks", ICCT, Beijing, China, September 1992.

[13] Guillemin F.M., Monin W., "Management of Cell Delay Variation in ATM Networks", GLOBECOM 1992, 128-132.

[14] Hübner F., "Dimensioning of a Peak Cell Rate Monitor Algorithm Using Discrete-Time Analysis", University of Würzburg, Institute of Computer Science Research Report Series, Report No.59, March 1993, to appear in Proc. 14-th International Teletraffic Congress, Antibes Juan-les-pins, France, June 1994.

[15] Hübner F., Tran-Gia P., "A Discrete-time Analysis of Cell Spacing in ATM Systems", University of Würzburg, Institute of Computer Science, Research Report Series, Report No. 66, June 1994.

[16] Hübner F., "Discrete-time Performance Analysis of finite-capacity Queueing Models for ATM Multiplexers", University of Würzburg, Institute of Computer Science, Ph.D. Dissertation, 1993.

[17] Hluchyi, M. G., Yin, N., "On the Queueing Behavior of Multiplexed Leaky Bucket Regulated Sources", INFOCOM 1993, paper 6a.3.

[18] Hunter J.J., "Mathematical Techniques of Applied Probability", Vol.1: Discrete Time Models: Basic Theory, Academic Press, 1983.

[19] Kingman J.F.C., "Inequalities in the Theory of Queues", J. Roy. Stat. Soc. B32(1970) 102-110.

[20] Kobayashi H., "Stochastic Modelling: Queueing Models; Discrete-Time Queueing Systems", in : Part II, Louchard G., Latouche G. (eds.), "Probability Theory and Computer Science", Academic Press 1983.

[21] Konheim A.G., "An Elementary Solution of the Queueing System GI/G/1", SIAM J. Comp., 4(1975) 540-545.

[22] Lindley D.V., "The Theory of Queues with a Single Server", Proc. of the Cambridge Philosophical Society, 48(1952) 277-289.

[23] Meisling T., "Discrete-Time Queueing Theory", Operations Research 6(1958) 96-105.

[24] Pack C.D., "The Output of an M/D/1 Queue", Operations Research 23(1975)4, 750-760.

[25] Reiss L.K., Merakos L.F., "Shaping of Virtual Path Traffic for ATM B-ISDN", INFOCOM 1993, paper 2a.4.

[26] Smith W.L., "On the Distribution of Queueing Times", Proc. of the Cambridge Philosophical Society, 49(1953) 449-461.

[27] Tran-Gia P., "Discrete-time analysis for the interdeparture distribution of GI/G/1 queues", Proc. Semin. Teletraffic Analysis and Comp. Perform. Eval., Amsterdam, The Netherlands, 1986.

[28] Tran-Gia P., "Analysis of a load-driven overload control mechanism in discrete-time domain", Proc. 12th International Teletraffic Congress, Torino 1988.

[29] Tran-Gia P., Ahmadi H., "Analysis of a Discrete-Time $G^{[x]}/D/1 - S$ Queueing System with Applications in Packet-Switching Systems", IN-FOCOM 1988, 861-870.

[30] Tran-Gia P., Dittmann R., "A discrete-time analysis of the cyclic reservation multiple access protocol", Performance Evaluation, Vol. 16, 1992, 185-200.

[31] Tran-Gia P., Rathgeb E., "Performance Analysis of Semidynamic Scheduling Strategies in Discrete-Time Domain", Proc. INFOCOM '87, San Francisco, March/April 1987, IEEE Computer Society Press 1987, 962-970.

[32] Wallmeier E., Worster T., "The Spacing Policer, an Algorithm for Efficient Peak Bit Rate Control in ATM Networks", Int. Switching Sympos. 14, October 1992, paper A5.5.

# 8

# PARALLEL PROCESSING OF PROTOCOLS

M. Björkman, P. Gunningberg

*Uppsala University, Uppsala and Swedish Institute of Computer Science, Stockholm-Kista, SWEDEN*

## ABSTRACT

Many workstations of today consists of a multiprocessor system with a shared memory and a small number of processors. We present a "processor-per-message" partitioning approach to do parallel processing of protocols. We have used the approach to implement a shared memory multiprocessor implementation of the x-kernel protocols and execution environment from University of Arizona. The two major performance bottlenecks in shared memory multiprocessor execution of protocols are lock contention and contention for shared memory. Measurement results on the implementation and simulation results of the locking effects are presented. The measured speed-up for the parallel implementation compared to the sequential one is more than 12 times for UDP and 3 times for TCP.

## 1  INTRODUCTION

The capacity of a host to process incoming and outgoing messages is a serious bottleneck in layered, high performance distributed systems. For a 100 Mbit per second network with 1 kbyte messages, two messages may arrive within 100 microseconds. For higher speed networks and smaller messages the interval will be even shorter. A regular workstation, running for example TCP/IP, can not process messages at this rate. There are several proposals on special hardware and software to reduce the processing time for a message. For a software alternative, see for example [1]. Parallel processing of protocols is an interesting alternative for multiprocessor systems. Having such a multiprocessor system, a number of the available processors may be dedicated to protocol processing.

We have chosen a parallelization paradigm which we will refer to as *processor-per-message*. It means that any processor in an array of processors can process the whole protocol stack for one message. An incoming or outgoing message is processed on the next available free processor and several processors may execute in parallel on different messages belonging to the same connection. Our aim

133

with this approach is to increase the throughput, the number of messages processed per second, rather than to decrease the latency, the processing time per message. The two main advantages with this approach is that it can support a small number of heavily used connections and that it has fairly coarse grained parallelism (message sized) compared to other alternatives, such as processor-per-function, e.g. [2]. When using small granularity there is a need to more frequently pass data and synchronization information between processors. A coarse parallelization will lower this overhead. For a longer discussion on different paradigms see the more extensive paper [3].

We have realized a multiprocessor protocol execution environment based on the processor-per-message paradigm. The environment is a parallelization of the x-kernel from University of Arizona which has a thread-per-message paradigm [4]. In this environment we have created parallel versions of TCP, UDP, IP and an Ethernet driver. The stacks TCP/IP/Ethernet and UDP/IP/Ethernet are two of the most used protocol stacks, and exhibit different behaviour. The parallel x-kernel is implemented as a user space server on a 26 processor Sequent Symmetry running Dynix V3.1.1.

In section 2, the effect of memory and lock contention on performance is discussed. Section 3 describes the parallel xkernel and in section 4 we present performance results of UDP/IP/Ethernet and TCP/IP/Ethernet and compare them to simulation results of lock contention effects.

## 2 LOCK AND MEMORY CONTENTION

In a system where several processors operate on the same connection or message, protocol state and data must be shared among the processors. Shared memory is an efficient way to handle the need for concurrent access to shared state and shared data. The two dominating performance affecting interaction factors for communication protocols in a shared-memory multiprocessor system are contention for the shared memory and software synchronization either because of execution ordering constraints, or because of accesses to shared state or data. For correct processing only one processor at a time may use a shared resource. Since processors may have to wait until shared resources are available, the parallel processing speed-up will not scale linearly. Locking is the normal way to ensure mutual exclusion in a shared memory system. When processors compete for locks, waiting occurs and this affects the performance. Locking limits are mainly dependent on the protocol implementation, and not on the machine it executes on. Machine contention (contention for shared resources such as buses or memory) on the other hand, is machine dependent.

The TCP protocol has a large state space and the performance of this stack is bound by TCP locking contention. UDP, on the other hand, has a much smaller state space which is also less frequently updated. The UDP performance is bound by machine contention rather than lock contention.

# 3  SPIN LOCKS

Some protocol state and data structures are shared among the processors, such as the sequence number in the window based TCP protocol. These state and data need to be shared exclusively in order to ensure consistency in the concurrent processing. A lock is granted to only one requesting processor at a time. It is typically a mutex variable which is guaranteed by the system to be updated atomically. A region of sequential code which contains updates to shared data structures is protected by locks. It could be the modification of a single variable, where the new value depends on the old one. It could also be larger updates, where the values of a set of variables depend on the values of another set of variables.

*Spin locks* are used in shared memory multiprocessor systems with local data and instruction caches. This means that waiting processors "spin" in a loop in which they continuously test the lock. There are two arguments for this kind of non-productive use of the processors. First, spinning processors do not disturb other processors because they are "spinning in the cache". Second, for short locking times it may be inefficient to do a time-consuming context switch to another process. We use spin locks in order to protect shared data, that is to ensure atomic updates. When a processor reaches a locked region it has to "spin" until the region becomes unlocked. Hence, during spinning the processor will not do any effective work and the effective parallelism is reduced. Locking can be costly if a protocol frequently accesses shared state and data.

The number and lengths of the critical sections put an upper limit on the achievable speed-up. For example, if the sum of critical sections are 20% of the total length, then the upper limit for the speed-up is five times, since only one processor at a time can hold the lock to this section.

Furthermore, with an increasing number of messages processed in parallel, the probability of spinning will also increase. We will present the effects of locking on protocol processing performance, when it will be the performance bottleneck in parallel protocol execution.

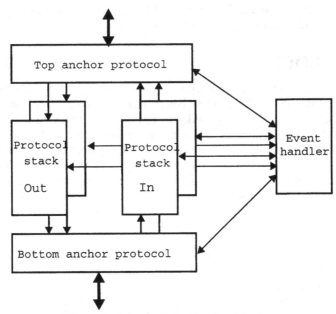

**Figure 1**    The parallel x-kernel architecture.

## 4  PARALLEL X-KERNEL

The (sequential) x-kernel is a protocol execution environment developed at University of Arizona [4]. It is a self-contained execution environment in which protocols can be implemented in a modular and uniform, yet efficient, way. The x-kernel is implemented on a number of different machines and operating systems. Both kernel and user space implementations are available.

In our *parallel* implementation, each processor is assigned a direction, inbound or outbound. A processor will execute the whole protocol stack when a message arrives or is ready to be send. (The reason for separate directions is to benefit from locality of reference when executing in the protocol code.) The top and bottom "anchor" protocols are implemented on separate processors as well as the event handler, see figure 1. The anchor protocols are due to the x-kernel way of structuring the interfaces to the user as well as to the underlying network. We use them in the measurements as sink and sources of messages. The event handler is very suitable for a dedicated processor. It handles a relative large common state space for the timer handler and for the threads/messages ready/blocked lists. If realized on an array processor these states must be frequently locked. It also does the time consuming interrupt handling which could be cumbersome if spread over several machines. At last there is very little interaction between the handler and its clients. Basically, the interface is schedule events, like set timer, reset timer, and message available for processing. In the figure, each box represents a processor.

The parallelization work was divided into two phases; parallelization of the x-kernel itself, and parallelization of the protocols TCP, UDP, IP and an x-kernel Ethernet anchor protocol. In the x-kernel itself, the three major tasks were parallelization of thread execution, memory handling and of event handling. We divide the different types of locks into three classes: *protocol specific*, *protocol implementation specific* and *environment specific*.

# 5  MEASUREMENT RESULTS

We measured the throughput with 1Kbyte user data messages on the parallel x-kernel for the protocol stacks UDP/IP and TCP/IP and an x-kernel Ethernet protocol. A high performance network driver was simulated by inserting a sink and a source in the Ethernet code. The sink flushes outgoing data, the source creates data and supplies it to the x-kernel as incoming frames. Apart from the sink and the source, the protocol implementations are interoperable with sequential TCP and UDP implementations. All measurements were done on our 26 processor Sequent Symmetry S81. Each processor is a 80386 CPU with 64 KB cache. A cache coherence protocol maintains consistent caches. The bus bandwidth is 80 MB/s. By using message sink and sources on the array processors we avoid copying messages over this slow bus, which considerably reduced the bus contention.

Figures 2 and 3 show the measurement results and with corresponding simulation results. The simulation results only consider the effect of locking contention not the bus contention. The simulations are based on traces from a single processor execution. A trace consists of time stamps of all the lock and unlock events for processing one message. Thereby we get the length and distribution of the locks and can simulate the locking effect, see [3] for a description of the simulation tool.

As can be seen in figure 2, the performance increase for TCP, outbound direction, starts to level off around 5 processors. Thereafter there is no substantial increase for additional processors. From the simulation, it is clear that locking put an absolute limit on the speed-up to about 4.5. The difference between the measured and simulated speed-up is due to machine contention. The speed-up for the inbound direction was very similar to the outbound direction and is therefore not included here. For UDP in figure 3 we did not see the corresponding level off for 20 processors in an array. This means that locking is not the limiting factors,

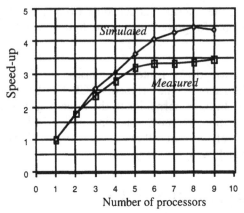

**Figure 2**    Comparisons of simulated and measured speed-up for TCP outbound

instead it is machine contention. For 20 processors the memory and bus contention of this machine limits the speed-up to 12 times.

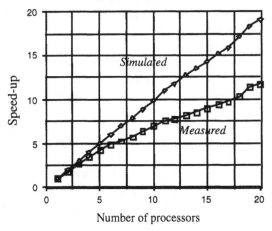

**Figure 3**    Comparisons of simulated and measured speed-up for UDP outbound

We believe that TCP could get a considerably higher speed-up if written directly for a parallel implementation. Now the TCP x-kernel implementation (which originates from the Unix BSD implementation) has unintentional large state space which is also modified all-over the code. Besides the unavoidable TCP state variables, like for the TCP sequence number, there are other implementation state variables introduced for the ease of implementation which can be avoided.

# 6 CONCLUSIONS

The protocol stacks UDP/IP and TCP/IP can execute efficiently in parallel on a shared-memory multiprocessor system. The measured speed-ups for the parallel implementations relative to the sequential implementations are more than 12 times for UDP and 3 times for TCP. These limits are set by spin locking to shared data structures and contention to the bus and memory. For protocols with small state space, as UDP and IP, the cost of locking shared state is small and is not the limitation of parallelism. For protocols with more complex state, as TCP, up to 5 processors can execute in parallel before locking saturates this particular system.

# REFERENCES

[1] Clark, D., Jacobson, V., Romkey, J., and Salwen, H., "An Analysis of TCP Processing Overhead", IEEE Communications Magazine, June 1989.

[2] Zitterbart, M., "High-Speed Protocol Implementations based on a Multiprocessor-Architecture", Proc. 1st IFIP Workshop on Protocols for High-Speed Networks, Zürich 1989.

[3] Björkman, M. and Gunningberg, P., "Locking Effects in Multiprocessor Implementation of Protocols", ACM SIGCOMM 93

[4] Hutchinson N. and Peterson L., "The x-kernel: An Architecture for Implementing Network Protocols", IEEE Trans. on Software Engineering, vol. 17 no 1, pp 64-75, Jan. 1991.

[5] M. Goldberg, G. Neufeld, M. Ito, "A Parallel Approach to OSI Connection-Oriented Protocols", Proc. 3rd IFIP Workshop on Protocols for High-Speed Networks, Stockholm 1992.

# 9

# AMTP: TOWARDS A HIGH PERFORMANCE AND CONFIGURABLE MULTIPEER TRANSFER SERVICE

Bernd Heinrichs

*Ericsson Eurolab Deutschland, Ericsson Allee 1*
*D-52134 Herzogenrath, Germany*

## ABSTRACT

This paper presents the service specification and an excerpt of a performance analysis of the Adaptive Multicast Transfer Protocol (AMTP) focusing on different error handling strategies for multicast communication scenarios. The need for sophisticated and dedicated multicast mechanisms to be provided by transfer protocols (covering protocols of layers 3 and 4 according to the ISO/OSI Reference Model) in order to support group communication becomes evident in the light of the versatile QOS (Quality of Service) requirements imposed by multimedia applications and the broad service spectrum offered by different network architectures. Today's protocols provide neither the required range of functionality nor the necessary performance.

Although various transport protocols are capable of providing some basic functionality essential for multicast applications, they do not address requirements for different delivery semantics, real-time support or dynamic group membership. However, AMTP is a first step in bridging the services gap by offering an application driven error control, mechanisms to reduce the acknowledgement implosion, support for different grades of reliability, guarantees for different QOS requirements, etc. in addition to usual transport layer protocol features. The configurability and feature richness of AMTP results from the call/connection principle, which splits a multimedia call into different media dependent connections, and a novel service concept, which subdivides the services allocated to a specific connection into primary and secondary ones.

## 1 PREAMBLE

AMTP builds the core of the DYCAT architecture [1] offering the service versatility and performance potential needed for configurable, application tailored communications. The modular design is one of the main characteristics of AMTP. It is derived from a formal protocol specification produced with higher level Petri Nets [2] and the subdivision of services into primary and secondary ones. The protocol and its formal description have been verified and different functionalities have been tested via simulation. Own experience shows that there is a discrepancy between theoretical design of new protocols and their practical realization. Thus future development should be accompanied by well-organized implementation and advance performance measurements in order to accelerate the acceptance of new concepts [4]. To take advantage of AMTP's performance gains a parallel implementation of the protocol is going on.

An overview of the protocol is given in Section 2 covering the description of the novel service concept based on the subdivision into primary and secondary services. The goal of this approach is to achieve an expedited protocol configuration and call/connection establishment prior to information exchange. The main features of the primary services, especially the multipoint related services, and their composition of secondary services are illustrated. Section 3 presents a more detailed description of AMTP's multicast protocol capabilities based on XTP [5] along with a description of performance results of selected protocol mechanisms in Section 4. Special emphasis is laid on the analysis of error control mechanisms for the different multicast services offered by AMTP. Compared to alternative approaches, one advantage of AMTP is the reduction of the number of acknowledgements sent in response to a sender's acknowledgement request. Although the number of CNTL packets will be reduced significantly, the resulting service may still be completely reliable (all-reliable if requested), as depicted in figure 1.

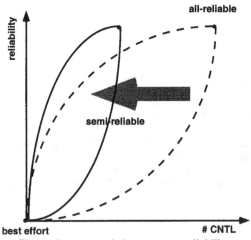

**Figure 1**    Acknowledgements vs. reliability

Error control on transfer level is of great importance especially on top of ATM based networks. Although very low error rates are being aimed for, the ATM layer makes no guarantee about the transport of individual cells [3]. A variety of mechanisms results in loss of cells across a virtual connection. This cell loss is an obvious problem for user services, as it forces loss of transfer or transport protocol data units and thus loss of user data. Since the ATM Adaptation Layers (AALs) only serve as error indicators and not as error recoverers the required functionality has to be added by the transfer systems. Since each point-to-point and multipoint transfer call may consist of several media dependent connections, the error handling is on connection level. Section 5 discusses the appropriate use of AMTP for different applications.

## 2 AMTP PROTOCOL SPECIFICS

The primary motivation for the design of AMTP is to prove that a single *monolithic* transfer protocol with a low complexity is capable of supporting a wide range of applications. To meet the varying performance and functionality requirements, the protocol is designed to be highly adaptive and configurable. In contrast to the conventional subdivision of transfer systems into two separate protocols, one on network, the other on transport layer, AMTP pursues the idea of a monolithic design as introduced in [5] with the exception of integrating routing functionality. Currently the link-state routing protocol OSPF (Open Shortest Path First [18]) is favourite, since it provides type-of-service routing and multicasting. At a first glance a monolithic design seems to be counterproductive to user tailored configurability. However, the use of fine grain service tuning facilities offered by AMTP achieve the opposite effect. As illustrated in figure 2, the user of a conventional transfer system has no chance to influence the processing of data in intermediate systems since the transport service interface is the end-to-end access point of users. Thus, users are kept from directly influencing and adjusting the network protocol in order to request an appropriate treatment of data in transit. The monolithic approach instead enables the user of the transfer system to specify and influence the processing of data in flow. Thus, AMTP is capable of guaranteeing different service quality grades.

Separation between calls and connections aims at supporting the transmission of multimedia data streams consisting of several different media by dedicated media dependent services. A call comprises a number of different connections having the same call end points but different service semantics. Data streams which request error correction and advanced QOS support will be supported by a respective connection oriented service whereas error and delay insensitive data may be carried by an unacknowledged connectionless service.

## 2.1 AMTP's Service Concept

AMTP's service concept is based on the distinction of primary into secondary services (cf. figure 3). Each primary service is characterized by default realizations of secondary services. The cooperation of the respective mechanisms determines the semantics of the primary service. If the transfer service user requests a

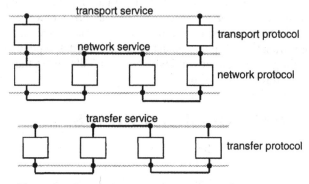

**Figure 2** Conventional vs. monolithic transfer systems

reconfiguration of the primary service, optional mechanisms are selected. The choice is limited in order to keep the semantics of the primary service and to support AMTP's fast configuration capability.

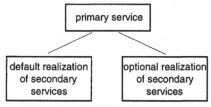

**Figure 3** AMTP service concept

AMTP provides eight primary services, split into point-to-point and multicast transfer services. Each end and intermediate system caches the mapping of an unambiguous pointer, called **context mask**, onto the related primary services. **context mask** is carried in AMTP headers. Using this service lookup concept speeds up service provision compared to nearly unrestricted configurable approaches [6,7,8,9,10]. Additionally, it reduces the number of control packets, which have to be exchanged during connection establishment, or during data transfer in order to initiate a service reconfiguration.

As depicted in figure 4, each AMTP packet comprises a **service**-field consisting of the **context mask**, the chmod-field (change mode) and a **chmod** indicator bit (CM). In addition to the format of the **service** field, the figure shows the values of **context mask** which have to be selected to activate one of the listed eight primary services. The values of <CM+**context mask**> are selected with the largest possible hamming distance, in order to reduce the probability of ambiguities resulting from bit errors. Change mode can be activated by the initiator, the

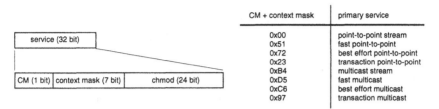

| CM + context mask | primary service |
|---|---|
| 0x00 | point-to-point stream |
| 0x51 | fast point-to-point |
| 0x72 | best effort point-to-point |
| 0x23 | transaction point-to-point |
| 0xB4 | multicast stream |
| 0xD5 | fast multicast |
| 0xC6 | best effort multicast |
| 0x97 | transaction multicast |

**Figure 4**   service field and context mask -> service mapping

receivers or any intermediate system involved in the respective connections in order to select optional secondary service realizations.

In contrast to alternative transport or transfer protocol architectures, AMTP offers a flexible packet format, which is selected according to the requested service. Otherwise, each packet header would have to cover the control information of all different services and service options. The packet header structure contains two parts, the four-byte AMTP Common Header and the variable length AMTP Packet Specific Header. Since it is beyond the scope of this paper the complete syntax of AMTP is not explained. The interested reader is referred to [17].

For none of the listed services (cf. figure 4) *multiplexing/demultiplexing* functionality is envisaged. Operating without this functionality enables a bijection between incoming and outgoing data streams at every processing entity. Additionally a dedicated support of different media belonging to the same application becomes viable. To do without multiplexing/demultiplexing enables AMTP to realize the concept of multimedia calls by supporting single media dependent connections with a dedicated quality of service.

Compared to the service definition of the OSI transport protocol, AMTP does not offer functionality for *splitting/recombining*. Splitting is not needed, since the separation into different media streams should have been done by more application related protocols [1].

## 2.2 AMTP Services

The eight different AMTP primary services are characterized by a number of adjustable secondary services. The user can select different mechanisms for the realization of the secondary services if the default mechanism is not her/his/its favored choice. AMTP offers different protocol mechanisms to realize the following secondary services:

- connection establishment and release
- flow, load and congestion control

- error recognition, notification and correction
- QOS semantics, QOS parameters and intermediate system support
- acknowledgement implosion (only for multicast services)

## Point-to-Point Services

Since it is beyond the scope of this paper to provide a detailled description of the four different point-to-point services offered by AMTP these services will only be touched on.

The _Point-to-Point Stream Transfer_ service (PPST) offers either a connection oriented duplex connection or a connection oriented simplex connection with acknowledgements. The quantitative QOS parameters transmission rate (segment size, requested rate, minimum rate) and transmission delay (maximum) are mandatory whereas the control of the tolerable error rate (maximum, consecutive) may be switched off. The default QOS semantics is compulsory, but optionally an adaptive service can be selected. Both QOS semantics require the monitoring of the QOS during the whole connection. In case of a compulsory service the connection will be released if the service quality degrades beyond a minimum acceptable value. In the adaptive case an unforced service degradation leads to a renegotiation of QOS values. In contrast to the "Peer-to-Peer Enhanced Connection-mode Transport Service" introduced in [11], PPST supports the service negotiation not only between end systems but integrates also the intermediate systems in the service renegotiation phase. This dedicated capability offers the possibility to guarantee a requested QOS. Additionally, PPST offers i.a. a "call blocking" mechanism used for limiting the number of simultaneously active connections in order to guarantee the QOS for existing connections.

The _Fast Point-to-Point Transfer_ service (FPPT) represents a connection oriented simplex data transmission without acknowledgements. Just a limited set of QOS parameters is available. Since transmission rate is not supported as QOS parameter, call blocking cannot be realized. FEC functionality (forward error correction) is the default error control mechanism because FPPT just offers a one way communication. Based on [12] the AMTP sender initiates a redundant number of packets, which is built via "logical XOR" of already sent packets [13].

The _Best effort Point-to-Point Transfer_ service (BPPT) provides connectionless simplex data transmission without or alternatively with ACKs. Intermediate systems are not involved in any service support. Thus, no QOS can be guaranteed. Checksumming as well as other possible error handling procedures can be switched off. Optionally to the default implicit connection establishment BPPT supports a "2 way handshake" in order to negotiate a sending rate. In that case, the sending rate is only kept by the source node.

The *Transaction Point-to-Point Transfer* service (TPPT) supports a request/ response semantics based on FPPT. Two connection oriented simplex connections will be established, which may have different QOS characteristics. Additionally, two joint QOS parameters concerning the response time and packet loss rate may be specified. To detect delayed or lost responses the initiator of a request uses timers. To compensate packet loss the XOR based FEC mechanism is used.

## Multicast Services

The primary services for the realization of multipoint communication are based on the above mentioned point-to-point services. The main difference is reflected in the error handling mechanisms. The multipoint services additionally offer a secondary service for the reduction of the acknowledgement implosion, taking place at the original initiator of the multicast association and at highly frequented intermediate systems. AMTP does not support full duplex connections between all members of a group. Only the initiator of a communication is able to send data in a multicast stream whereas the other group members are just allowed to react via unicast connections to the single initiator.

As representative of the multicast primary services the *Multicast Stream Transfer* service (MST) will be described most detailed. Similar to the PPST service MST offers simplex as well as duplex connections. The QOS will be negotiated and guaranteed with the assistance of intermediate nodes. An additional task of the intermediate nodes is to reduce the number of redundant acknowledgement packets flooding the network. This option may be switched off and substituted by the end system based damping and slotting strategies known from XTP [5]. The XTP option is only useful in local environments and may be selected by setting the XTPLIKE flag in chmod. The following table 1 summarizes the default and optional realizations of the secondary services associated to the MST service. The terms in brackets represent the flag to be set in chmod in order to activate the optional realizations of the respective secondary service. A detailed description of the different secondary service realizations is provided in [17].

The error handling routines offered by the MST service are similar to those offered by PPST. The default realization comprises header and data checksumming, sender based error control requesting acknowledgements from the receivers along with compound selective retransmissions. Optionally, intermediate nodes may be triggered to become responsible for retransmissions. This functionality is used to reduce network load, to relieve the initiator from retransmission handling and to achieve faster retransmissions. To prevent an acknowledgment packet from traversing the whole network up to the initiator of the call, the packets will be time stamped with a small "time to live" value. Thus, the packets are just able to reach the closest intermediate node before the time stamp expires. This functionality is activated by setting ROUTRET in chmod. Alternatively, a

| secondary service | default realization | optional realization |
|---|---|---|
| connection establishment | 2 way handshake with integrated key and service exchange (simplex) | 3 way handshake with integrated key and service exchange (duplex) (CONHS3) |
| flow control | rate based with an overlaid time based window mechanism (dynamic) | deactivation of the window mechanism (NOWND) |
| congestion control | FF-Tri-S (incl. alarm messages) | deactivation of FF-Tri-S (NOFFTriS) |
| error recognition | header and data checksum, sender initiated, spans in receiving data stream are detected | deactivation of checksumming (NOCHECK); not sender based anymore (XORFEC) |
| error notification | compound selective acknowledgements, high priority CNTLs | deactivation of selective acknowledgements (NOSACK) -> positive CNTLs) |
| acknowledgement implosion | topology dependent filtering supporting different reliability grades | end system based damping and slotting for local areas (XTPLIKE) |
| error correction | time based sender initiated ACK-requests; block selective retransmissions | FEC based on XOR (XORFEC) router based retransmission (ROUTRET) |
| quantitative QOS parameters | transmission rate, transmission delay, error rate | deactivation of error rate monitoring (NOQOS3) |
| QOS semantics | compulsory | adaptive (QOSADAP) |
| intermediate system support | static priority queues, deadline scheduling, call blocking based on "requested transmission rate" | deactivation of static priority queues (NOSP); deactivation of call blocking (NOCB) |
| connection release | implicit (timer based) | 3 way handshake (DISHS3) |

**Table 1**    MST secondary services

forward error correction mechanism can be selected, based on XORing transmitted data. Figure 5 illustrates the DATA and CNTL packet flow for the simplex and duplex MST service.

The remaining multicast services are counterparts to the respective point-to-point services. The _Fast Multicast Transfer_ service (FMT) corresponds to the FPPT service, the _Best effort Multicast Transfer_ service (BMT) to BPPT and the _Transaction Multicast Transfer_ service (TMT) to TPPT.

# 3  AMTP MULTICAST

Selecting one of the multicast related primary services by adjusting the context mask enables an AMTP entity to provide multicast communication. Compared to XTP's semi-reliable approach AMTP provides a much broader range of service types, although AMTP's multicast flow and error handling is mainly based on XTP's bucket algorithm. The basics of the bucket algorithm are illustrated in the following subsection. AMTP extensions are described thereafter.

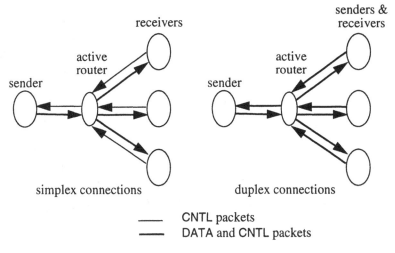

**Figure 5**  MST connection types

# 3.1 Multicast Error Handling Basics

The bucket algorithm defines a rate based flow control for multicast connections which is overlaid by a time based window control. The time windows together with the collection of control information received in the respective time frames are called *buckets*. Both, rate and window control may be switched off. Since the algorithm additionally triggers the sending of acknowledgement requests, it is responsible for error handling. Thus, to achieve semi-reliable semantics the flow control should be kept active. The original algorithm is solely developed for use on top of a broadcast transmission medium. Since the sender is not able to identify the receivers, the algorithm is unreliable or at most semi-reliable.

The receivers are passive, they only respond when being prompted. As with unicast, the multicast initiator sets the Status REQuest bit (SREQ) in the ptype field to request receivers to issue a control packet (CNTL). Each time, the multicast initiator generates an SREQ, a bucket is created. The time a request was issued, is recorded at the transmitter using the synch counter. Synch counter will be incremented by one each time a DATA packet is sent. The incoming CNTLs initiated by the receivers contain echo values that reflect the receivers' values of sync. This mechanism enables the sender to sort incoming CNTLs by 'age', i.e. the sync value. CNTLs in the same bucket are coalesced, recording the minimum of the flow parameters, the byte-based sequence number (rseq) indicating the highest consecutive sequence number received, and the maximum of the round trip times to the different receivers.

To reduce multicast overhead, the number of redundant CNTLs generated in response to an SREQ has to be decreased by efficient acknowledgement mechanisms and buffer control strategies. A CNTL can either cover a REJECT, indicating the first packet lost, or it can transport a SELective ACKnowledgement. With XTP, error control in multicast mode is based on REJECTs with a consecutive go-back-n retransmission. CNTLs are not exclusively addressed to the multicast sender but to the whole group. Consequently, every group member receives the packet and skips its own retransmit request if incoming CNTLs request retransmission of packets with the same or a smaller sequence number. This strategy is called <u>DAMPING</u>. The main disadvantage of the proposed damping mechanism results from its fragility due to timing considerations. To make damping feasible, each receiver lets expire a random delay prior to sending her/his/its own CNTL (<u>SLOTTING</u>). Obviously, there is a trade-off between efficiency and reliability which will be eliminated by AMTP. The more reliable the multicast transmission, the less efficient it will be.

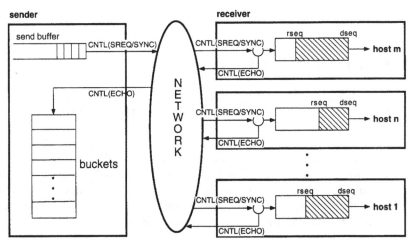

**Figure 6**   Model of the bucket algorithm

Figure 6 depicts a simplified queueing model of the communication scenario, valid for XTP as well as for AMTP, illustrating the packet exchange between sender and receivers and covering the relevant acknowledgement numbers, rseq and dseq.

Since XTP supports only go-back-n retransmissions, rseq is the basis for deciding on packets to be retransmitted. Compared to rseq, dseq marks the sequence number of the last byte, which has been processed completely and passed to the application by the receiver.

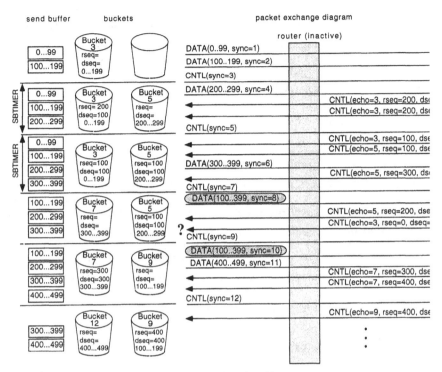

**Figure 7** XTP bucket algorithm

Besides the sequence numbers and the echo field covering the copied sync value, the sender uses the techo values of the incoming CNTLs to actualize the round trip time. Since the CNTLs will arrive at different times, the evaluation process must be delayed a specific time frame. This time frame corresponds to the life time of a bucket. Since the receivers may be dispersed all around the underlying network, the values of the buckets' life times are the critical parameters for the efficiency and reliability of the multicast connection. The life time depends on the number of initiated buckets per round trip time (SREQS), the actual round trip time itself (WTIMER), and an error tolerance factor DROPS. The Switched Bucket Timer (SBTIMER) which instantiates the release of obsolete buckets and the establishment of new ones is calculated as WTIMER/ (SREQS+0.5). How these parameters influence the efficiency and reliability of the algorithm is described in [5].

To illustrate some of the deficiencies of the bucket algorithm (transmission of redundant DATA or unnecessary CNTL packets) the following figure shows an example packet flow. As proposed in [5], there is no activity in the routers. The packet exchange diagram comprises four components. The send and the activated buckets represent the initiating part of the multicast connection. The send buffers store all transmitted DATA packets, held no longer than one bucket life time. The value of SREQS in the depicted example is 1. That means, at most two buckets are active simultaneously. Each bucket is characterized by its sync value in order to map the incoming CNTLs to the appropriate bucket. sync is incremented for each outgoing DATA and CNTL. The sequence numbers depicted at the bottom of the buckets represent the sequence numbers of the data, which has been transmitted in the respective SBTIMER interval.

The receiver group is connected to the sender by means of a packet exchange diagram. Each time the initiator generates a CNTL, which requests an acknowledgement from the receivers, a new bucket is created. When e.g. creating the bucket with sync 7, bucket 3 has to be released. All DATA with sequence numbers higher than the coalesced rseq will be retransmitted. This results in a repeated transmission of the same data. As represented in the above figure, data with sequence numbers 300 to 399 are transmitted three times within one round trip time. Another drawback is the huge amount of CNTLs transmitted in response of an ACK request. If the router would be an active transfer router, it would be able to filter CNTLs which need not to pass through, since a preceded CNTL already covered the same or a larger retransmission amount.

## 3.2  AMTP Extensions

The basic differences between AMTP's multicast mechanism and XTP's bucket algorithm are summarized in the following table 2. AMTP offers various strategies to reduce the ACK implosion at the sender. On the one hand the involved

| XTP | AMTP |
|---|---|
| semi-reliable | semi- and all-reliable |
| ACKs to the whole group | ACKs to the connection initiator |
| ACK implosion<br>• damping and slotting in<br>  end systems | ACK implosion<br>• filtering in intermediate systems<br>• group acknowledgement<br>• distributed acknowledgement |
| go-back-n retransmission | selective retransmission |
| constant transmission rate | rate adaptation based on acknowledgements |
| retransmissions from connection initiator | retransmissions from connection initiator and routers |
| anonymous receivers | concatenation of address information |
| bucket algorithm for error control in LANs | enhanced bucket algorithm for error control in WANs |

**Table 2**   XTP vs. AMTP multicast features

routers play an active role in the reduction of acknowledgement packets. The number of **CNTLs** traversing through the network to the initiator will be decreased at every intermediate system (**active router** approach). The alternative approach tries to reduce this number as close as possible to the receivers (**subgroup** approach).

To realize the **subgroup** approach two different strategies are feasible. When establishing the **group acknowledgement** strategy as depicted in figure 8,

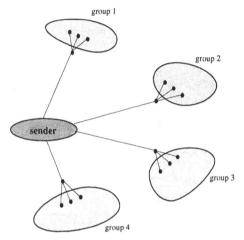

**Figure 8**   Group Acknowledgement

receivers at the same subnetwork form a subgroup. The acknowledgements sent are collected in the closest router to the respective subnetwork and forwarded in a single composite control packet. To achieve this functionality a collector just needs to set up its address filters appropriately and needs some logic to coalesce control packets. Using this strategy does not restrict the feasible grades of reliability.

Since AMTP offers a compound ACK packet, the routers can put the addresses of all reacting receivers into the respective packet header. Thus, full reliability with a significantly lower number of **CNTLs** becomes quite possible.

Alternatively, AMTP offers the **distributed acknowledgement** strategy, which is not designed to achieve all-reliability but high performance. Figure 9 illustrates the respective scheme, where the intermediate systems do not need to be active. The distributed scheme is based on the assumption that stations at the same local subnetwork suffer from nearly the same problems. Thus, the retransmission requests of those stations would overlap each other and it would be sufficient, if only one station responds to the sender's acknowledgement request.

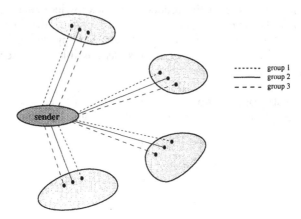

**Figure 9**   Distributed Acknowledgement

Each time the initiator requests an acknowledgement, she/he/it addresses a differ-ent subgroup of the receiver set with at least one station per subnet in each sub-group.

Besides the large number of **CNTLs** another serious drawback of XTP's multi-cast mechanism is the huge amount of superfluous data retransmitted at the release times of buckets. To reduce or even to remove all redundant retransmis-sions the following two algorithms are offered by AMTP.

The **retransmission cut** method (cf. figure 10) retransmits data with sequence numbers less than the highest sequence number sent in the respective bucket sending period (**SBTIMER** interval). Thus, the retransmission depicted can be reduced from the sequence number span 100...399 (cf. figure 7) to 100...299. To execute this filtering mechanism, just a minor change has to be undertaken to the basic XTP algorithm. Besides **rseq** the initiator has to keep in mind the highest

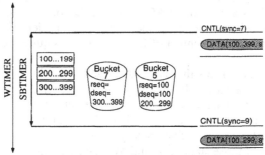

**Figure 10**   Retransmission Cut

sequence number sent during bucket sending time. Thus, not all packets with sequence numbers exceeding **rseq** will be retransmitted.

Using the second alternative, the **WTIMER Period** strategy, any redundant retransmissions per round trip time will be omitted. Thus, subsequent retransmissions of the same **DATA** packets are skipped. To realize the algorithm the initiator just has to stamp the sent data which are stored in the send buffers, with the time for the earliest possible retransmission. Prior to expiration of the respective timer no repeated retransmission is allowed.

# 4 AMTP PERFORMANCE ANALYSIS

To verify AMTP's multicast capabilities the respective algorithms have been specified using SDL [14]. The SDT tool [15] was used for the graphical representation of SDL constructs. The subsequent conversion into a simulation program enabled the detailed performance evaluation of AMTP. The following network topology (cf. figure 11) has been used for testing most of AMTP's multicast features.

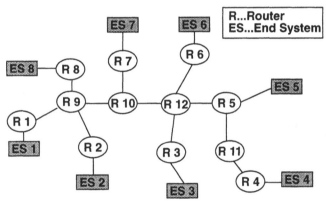

**Figure 11**   Simulation test scenario

The transmission of data was assumed to be bit-errorfree, which is a justified assumption at a bit error rate of $10^{-12}$ on fiber optic networks. The following basic input parameters have been selected for all the experiments:

- amount of transmitted data per multicast connection: 3 Mbyte
- packet length: 1024 bytes
- send and receive buffer lengths: 64 Kbyte
- transmission rate per end system: 50 Mbps
- medium transmission capacity: 100 Mbps

- packet sojourn time per router: 1 ms
- propagation speed: $2 \times 10^8$ m/s
- distance between systems: 1000 m

The performance results presented in the following subsections mainly cover the delay metrics achieved by transmitting data from end system 1 to all other end systems depicted in the above figure. The length of **SBTIMER** is adjusted according to the respective **WTIMER** interval and an **SREQS** value of 2.

## Unicast vs. Multicast Acknowledgements

The first experiment covers the comparison between different strategies to reduce the number of **CNTLs** reaching the initiator during a multicast connection (cf. figure 12).

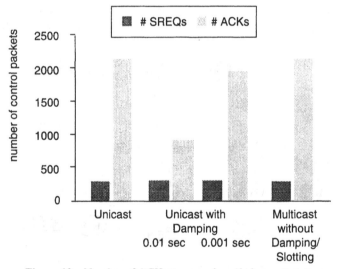

**Figure 12**   Number of ACKs versus acknowledgement strategy

Since XTP's damping and slotting only works in LANs, both mechanisms are switched off. Thus, as depicted in figure 12, the use of the "Multicast" and the pure "Unicast" approach result in nearly the same number of **CNTLs** reaching the initiator of the data transmission. Compared to those values the execution of damping (filtering) in routers leads to a significant reduction of the respective acknowledgement numbers. A damping time of 1 msec results in a 10 % reduction of **CNTLs** whereas a 10 msec damping time achieves a 60 % reduction. These figures motivate the involvement of routers in the reduction of acknowledgement traffic, especially in multicast communication scenarios.

The drawback of such algorithms is the inherent increase of total transmission time which results from the additional sojourn times in the routers. As illustrated in figure 13, the CNTL packet reduction of nearly 60 % in case of 10 msec damp-

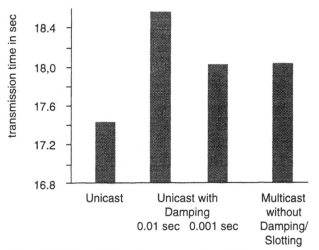

**Figure 13**   Transmission time versus acknowledgement strategy

ing times (cf. figure 12) is achieved at the expense of an 8 % extension of the total transmission delay.

A further improvement of AMTP's filtering algorithm may be achieved by using more dynamic mechanisms. To avoid an untolerable increase of waiting times, the sum of all damping times in a communication path should be not significantly longer than the maximum round trip time for a point-to-point connection. For the above mentioned strategies this means, the damping time per router would be so small, that a probable filtering is extremely unrealistic. Thus, instead of common static and equal damping intervals, different criteria are proposed to subdivide a total damping time onto the routers.

## Dynamic vs. Static Damping Intervals

In the following experiment two new approaches are compared. First, the longest damping time has been adjusted to the router closest to the initiator (**sender distance**). The obvious drawback of this approach is the reduction of load only close to the initiator, whereas the network condition remains unchanged.

An alternative strategy is to make the filtering times **topology** dependent. That is, a router with a large number of incident routers should be assigned a longer slotting interval than a router with a small number of neighbour routers [16]. In order

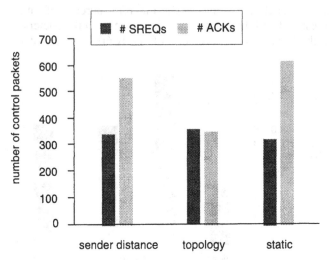

**Figure 14**   Transmission times versus damping strategy

to make use of a **topology** dependent algorithm, the network topology must be known to each router. Link state algorithms, for instance, provide every router with information on the complete network topology. For distance vector routing protocols the routers have to exchange configuration information at connection establishment. A possible allocation of damping times is to adjust no damping time to routers with the minimum number of neighbours in the respective multi-cast connection, whereas routers with the largest number of neighbours will achieve the longest damping times.

The examination of these algorithms show, that the topology based approach achieves a significant reduction of CNTLs compared to the static algorithm, whereas the sender distance approach is not very successful.

## 5   CONCLUSION AND FURTHER WORK

Based on the service specification, it is obvious that the main benefit of using AMTP as transfer system is the concept of predefined services. Although eight different services are offered to the user the appropriate service creation and the respective protocol configuration can be executed very fast using the context mask concept. The high flexibility, high performance and the provision of new multipoint services, makes AMTP an attractive protocol for supporting applications with changing and high performance demands.

Unfortunately, it was not possible to describe AMTP's protocol functionalities in more detail. The interested reader is referred to [17], where an extensive design description and a performance evaluation of AMTP capabilities such as real-time support or congestion avoidance are presented. Additionally, implementation basics are described.

Currently, the suitability of AMTP as an enhanced ATM adaptation layer is evaluated. The flexible service concept enabling the configuration of a "thin" or "thick" functionality protocol makes AMTP a promising candidate for an integrated approach, eliminating the need for additional transport functionality on top of it.

# REFERENCES

[1] Heinrichs, B., "DYCAT: A Flexible Transport System Architecture", Proc. IEEE International Conference on Communications ICC'93, pp. 1331-1335, Geneva/Switzerland, May 23-26, 1993

[2] Burkhardt, H.J., Ochsenschläger, P., Prinoth, R., "Product Nets - A Formal Description Technique for Cooperating Systems", GMD-Studies No.165, Gesellschaft für Mathematik und Datenverarbeitung mbH Bonn, December 1989

[3] Partridge, C: "Gigabit Networking", Addison-Wesley Professional Computing Series, 1993

[4] Heinrichs, B., Meuser, T., Spaniol, O., "High Speed Interconnection of Workstations: Concepts, Problems and Experiences", Proc. IFIP International Conference on Decentralized and Distributed Systems, Spain, September 1993

[5] "XTP Protocol Definition, Revision 3.6", Protocol Engine Incorporated, January 1992

[6] Haas, Z., "A Protocol Structure for High-Speed Communciation over Broadband ISDN", IEEE Transactions on Communications, vol. 38, no. 9, pp. 1557-1568, September 1991

[7] Plagemann, T., Plattner, B., Vogt, M., Walter, T., "A Model for Dynamic Configuration of Light-Weight Protocols", Proc. 3rd IEEE Workshop on Future Trends of Distributed Systems, 1992

[8] Schmidt, D., Box, D.F., Suda, T., "ADAPTIVE: An Object-Oriented Framework for Flexible and Adaptive Communication Protocols", Proc. 4th IFIP Conference on High Performance Networking, Liege, Belgium, 1992

[9] Zitterbart, M., Stiller, B., Tantawy, A.N., "Application-Driven Flexible Protocol Configuration", Proc. ITG/GI-Fachtagung "Kommunikation in Verteilten Systemen", pp. 384-398, 1993

[10] Hoschka, P., "Towards tailoring protocols to application specific requirements", Proc. INFOCOM'93, pp. 647-653, 1993

[11] ISO/IEC JTC1/SC6/WG4 N822, "FOR INFORMATION - Enhanced Transport Service Definition (Informal Specification in English - Version 1)", April 1993

[12] Shacham, N., Mc Kenney, P., "Packet Recovery in High-Speed Networks Using Coding and Buffer Management", Proc. IEEE INFOCOM'90, pp. 124-131, 1990

[13] Aghadavoodi Jolfaei, M., Heinrichs, B., Nazeman, M., "TCP Extensions for Interconnection of LANs by Satellite", Proc. IEEE National Telesystems Conference (INDC) '94, Madeira, Portugal, April 1994

[14] CCITT Recommendation Z.100, "Specification and Description Lanuage SDL", Contribution X-R-15-E, 1987

[15] "SDT 2.2 Reference Manual Volume 1", Telelogic, 1992

[16] Fichtner, M., Heinrichs, B., Jakobs, K., "A Transfer System for Multi-Media Group Communication", 3rd IEEE Workshop on Enabling Technologies - Infrastructure for Collaborative Enterprises, 1994

[17] Heinrichs, B., "Transfersysteme zur Hochleistungskommunikation", PhD Thesis (Technical University of Aachen), in German, 1994

[18] Moy, J., "OSPF Version 2", RFC 1247, July 1991

# AN INTERNETWORKING ARCHITECTURE FOR MULTIMEDIA COMMUNICATION OVER HETEROGENEOUS NETWORKS

M. Graf, H. J. Stüttgen

*IBM European Networking Center*
*P.O. Box 10 30 68*
*D-69020 Heidelberg, Germany*

## ABSTRACT

Distributed multimedia applications of the future will run on heterogeneous networks. Initially deployed on traditional LANs like Token Ring and FDDI there will be a long period of transition to ATM-based LANs that offer better support for continuous media traffic like video and audio. In parallel, public ATM services emerge that enable the wide-area interconnection of LANs. In this paper we identify the issues involved in the interconnection of heterogeneous networks for multimedia communication. We discuss the interaction of traffic management, signalling and data transmission protocols and present alternatives in the protocol architectures.

## 1  MOTIVATION

Distributed multimedia applications like workstation conferencing, multimedia sales catalogs, distributed learning or interactive TV require the integration of audio and video data with non realtime data in the communication system. This integration can be achieved on two different levels. In the continuous bit stream oriented (CBO) approach the audio and video data are kept separate from the non realtime data and only integrated at the presentation level by means of video overlay boards. A much more flexible approach is to handle realtime streams like normal computer data, by transforming the coninuous media stream in a packetized data stream. Here we can take advantage of the existing infrastructure, from storage devices to communication networks. In this paper we will focus on the packetized approach.

In order to utilize existing networks for packetized multimedia communication, a communication architecture supporting the realtime characteristics of AV streams is needed. To avoid the introduction of transmission delays due to network contention into the delay sensitive AV streams, it is mandatory to reserve the necessary resources like network bandwidth or buffer space along the path of the data. Thus a suitable communication architecture requires the use of reservation protocols. Reservation protocols need to be able to carry the resource requirements of a stream along the desired network route and negotiate provision of the respective resources with the resource management entities. One example of a resource management protocol is the ATM Q.2931 signalling protocol. Q.2931 allows the user to specify his traffic requirements and communicate them to the ATM network. However, Q.2931 is targeted solely at a pure ATM environment e.g. it is not applicable to other networks like existing LANs. Given the fact that our current networking environment consists today of, and will include for the foreseeable future a variety of LANs like Ethernet, Token Ring and FDDI a more general approach is required. In the environment of heterogeneous interconnected networks network layer reservation protocols, which can negotiate resource reservation across subnetwork boundaries, have been developed.

Currently two different approaches are under discussion. The first of which is the Internet Stream Protocol Version 2, ST-II [1]. Different implementations of ST-II are under way. One as part of the ENC's HeiTS multimedia stack, others within the German BERKOM project, where ST-II is the basis for the BERKOM multimedia transport system, which is being developed for DEC, HP, IBM, Siemens/ SNI and SUN workstation platforms. At the same time a working group of the IETF is reworking the ST-II specification in order to reflect the initial development experience. Alternatively a new protocol called RSVP [4] has been proposed to enhance the ST-II approach by features as receiver oriented multicast control and heterogeneous multicast trees. Although we will focus in our discussion below on ST-II, similar argumentation applies to RSVP. However, as a detailed RSVP specification has not been published, RSVP can at this time not be used for a detailed analysis.

Besides the reservation protocol that is needed to communicate traffic requirements between the different entities in the various subnetworks, another issue is the description of stream characteristics. Whereas a video application may be able to specify a required frame rate to the communication system, an ATM network expects a specification based on ATM cell rates. Not only is the relationship between frame rate and cell rate for variable bit rate streams not a straight forward calculation, but also neither the ATM cell rate nor the video frame rate may be a useful traffic description for LANs. Thus the issue of traffic description and traffic management based on the provided traffic description needs to be covered independently of the respective reservation protocol.

In this paper we will investigate an interworking architecture that is able to cover the resource reservation problem in the environment of heterogeneous subnetworks. In particular we want to include the existing LAN base and evolving ATM networks. The following section will first introduce the existing background, namely the ST-II approach for packet networks and ATM. In section 3 we will study the different interworking problem areas which are traffic management, control (signalling) protocols and data interworking individually. Finally we will outline how the described architecture will be implemented as part of the BERKOM B-ISDN pilot project.

# 2 BACKGROUND

## 2.1 ST-II

Internet Stream Protocol Version 2 (ST-II [1]) is a connection-oriented network layer[1] protocol. It is intended to interconnect heterogeneous subnetworks and is thus located at the same layer as the connectionless Internet Protocol (IP). It consists of two parts: the Stream Protocol (ST), a lightweight protocol carrying data packets, and the Stream Control Message Protocol (SCMP) for connection control (see Figure 1). Like IP, ST-II expects from the subnets an unreliable datagram service.

The ST protocol is a lightweight store-and-forward protocol. The packet header is simplified in comparison to IP. Instead of carrying the sender's and receiver's IP address, a short label, called *Hop Identifier* (HID) identifies the connection to which a packet belongs. An *ST Agent* (an ST-II protocol instance on an end system or router) receiving a packet uses the HID to access its internal data structures containing the connection's state information to determine the destination, e.g., the outgoing network interface. HIDs are unique on a *Hop*, i.e. a subnet between two ST Agents, and they are overwritten by each ST Agent. The HID values are assigned at connection setup time. In summary ST-II uses a label-switching technique on the packet level.

SCMP takes up the main part of the ST-II specification. ST Agents exchange control messages to set up and release connections, to negotiate the traffic contract and to change the connection topology. SCMP messages are encapsulated in ST messages and are identified by a reserved HID value of 0. In order to guarantee reliable transactions on top of an unreliable datagram service, SCMP messages are structured according to the request-response principle: each SCMP message must be acknowledged by a corresponding response message. Unacknowledged messages are retransmitted after time-out.

ST-II assumes that the packet switching using the ST protocol is controlled by real-time mechanisms, like, e.g., a packet scheduler [8], in order have a certain QoS guarantee. Separate resource administrators are necessary that control node

---

1. in OSI terminology

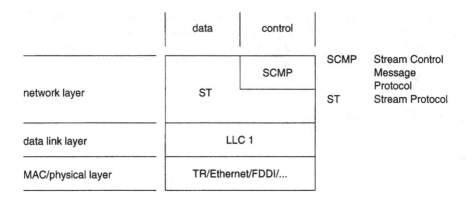

**Figure 1**   ST-II protocol stack in a traditional LAN

and network resources. The ST-II specification does not specify these mechanisms and it makes only a proposal for the quality of service and traffic parameters contained in the SCMP messages. ST-II has mechanisms to communicate
and negotiate the traffic contract, but it is up to the individual implementation to
determine the parameters that fit the chosen resource allocation architecture.

ST-II offers to the higher layers a connection-oriented service with unidirectional
point-to-multipoint connections. Optionally a one-step setup of bidirectional
point-to-point connections that logically consist of two independent unidirectional connections is provided. At connection setup time a traffic contract negotiation service is provided between the sender, the receiver(s) and the network.
Receivers can be added to and removed from an already existing connection.

## 2.2  ATM

ATM [2], [3] as standardized by ITU follows a layered approach (see Figure 2).
Above the Physical Layer the ATM layer transmits and switches *cells* that consist
of a 5-byte header and a 48-byte payload field. The header contains a label that
identifies the connection a cell belongs to. The label is split in a Virtual Circuit
Identifier (VCI) and Virtual Path Identifier (VPI) forming a two-level hierarchy
of connections. The ATM Layer offers a connection-oriented service based on
cells. ATM Layer connections are unreliable (cell loss, corruption, misinsertion),
but have some performance guarantee both concerning the degree of reliability
and the temporal behavior, depending on the negotiated Quality of Service.

The ATM Adaptation Layer (AAL) at the edge of the ATM network adapts the
service of the ATM Layer to the specific needs of higher layer protocols. These
needs are indeed very differing depending on the use of the AAL connection. A

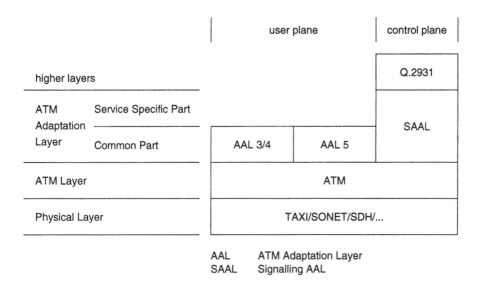

Figure 2 shows the ATM protocol stack. The layout:

| | user plane | | control plane |

| higher layers | | | Q.2931 |
| ATM Adaptation Layer — Service Specific Part — Common Part | AAL 3/4 | AAL 5 | SAAL |
| ATM Layer | ATM | | |
| Physical Layer | TAXI/SONET/SDH/... | | |

AAL     ATM Adaptation Layer
SAAL   Signalling AAL

**Figure 2** ATM protocol stack for packet-oriented service

stream of packetized continuous-media data will want a lightweight framing protocol preserving the temporal Quality of Service of the ATM Layer connection, whereas discrete media will request a service that is only best-effort regarding time but made reliable by the AAL.

For packetized data, two AAL protocols are currently defined: AAL 3/4 and AAL 5. The ITU has divided the AAL into nested sublayers. The Common Part (CP) of AAL 3/4 and AAL 5 offers a packet-oriented service, by segmenting packets into cells and reassembling cells into packets. AAL 5 only performs framing and checksum calculation whereas AAL 3/4 additionally enables the interleaving of packets in the cell stream. The Service Specific Part is located on top of the Common Part. For connections carrying continuous media it could be empty, i.e. higher layers directly use the service of the Common Part. For higher layers needing a reliable service, like, e.g., the signalling protocol, the Service Specific Part contains a protocol like the Service Specific Connection Oriented Protocol (SSCOP) that uses sequence numbers and retransmission for reliability.

In its B-ISDN Reference Model, the ITU distinguishes between a user plane and a control plane. The user plane contains the protocol stacks for data transmission while the control plane contains the stack for signalling. Signalling is done out-of-band, i.e. ATM connections separate from the data connections are used. The signalling protocol for B-ISDN will be specified in recommendation Q.2931, formerly Q.93B [5]. Additionally the ATM Forum has defined its own version of

Q.2931 [6]. Q.2931 uses the reliable service of the Signalling AAL (SAAL) that consists of SSCOP on top of the Common Part of AAL 5, thus Q.2931 itself has no mechanisms to ensure reliability.

As Q.2931 is in the process of being standardized, there is no stable version yet. ATM Forum is generally ahead of ITU regarding features of Q.2931, so we will use the current Forum version as basis in the following discussion of the ATM service. It offers unidirectional point-to-multipoint and bidirectional point-to-point connections. The parameters of the traffic contract are cell-oriented, like peak and mean cell rate. Currently the traffic contract cannot be negotiated (work item for next version). Like in ST-II, receivers can be added to and removed from an existing connection.

## 3  INTERWORKING PROBLEM

In the following we will focus on interworking issues in three areas: traffic management, signalling and data transfer.

## 3.1  Traffic Management

Traffic management is concerned with the provision of a certain level of quality of service in the data transfer phase, including packet delay and loss objectives. There are a variety of proposals for traffic management schemes for different network technologies in the literature, including packet switches for WANs [7], shared-medium LANs [8], and ATM [9] (for an overview see [10]). The common approach of these schemes is to make performance guarantees on a per-connection basis by reserving resources individually for each connection.

The main components that constitute a traffic management architecture are a policing function at the entrance of the network that limits the incoming traffic to a prespecified amount, an admission control function that determines whether a new connection can be accepted, a scheduling algorithm that determines the order in which to forward packets and a buffer management scheme that decides which packets must be dropped. When we consider a heterogeneous subnets with specific traffic management schemes, each of these components could be different. It is therefore necessary to have an abstract view of traffic management.

All traffic management schemes implicitly or explicitly embody the concept of a *traffic contract* between user and network. The traffic contract is an agreement that is closed on a per-connection basis. The user agrees to bound the traffic he generates to a prespecified amount. In turn the network, knowing the bounds imposed on all sources, is able to deliver the traffic with some quality of service guarantees. Therefore the traffic contract consists of two parts:

■  a source traffic description.

■  a quality of service description

For dynamic connections the traffic contract is closed at connection setup time and may be negotiated between user and network (for example in ST-II). For this negotiation user and network exchange a set of parameters that quantify the quality of service and the source traffic. In ST-II this set is called FlowSpec and is part of the SCMP messages, in ATM the parameters are part of Q.2931 Information Elements.

Figure 3 shows the approach adopted for an end-to-end connection in concatenated heterogeneous subnetworks. The subnetworks are subcontractors to deliver the end-to-end traffic contract. Each subnet contributes its share to the end-to-end quality of service, which is the aggregation of the individual quality of service guarantees. The traffic passing through the subnetworks is obviously everywhere the same, although it may be transported on different levels: at the packet level for packet-switching networks or on the cell level for cell-switching networks.

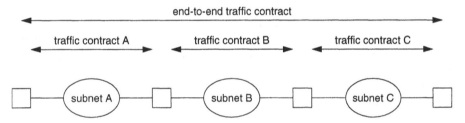

**Figure 3**   Concatenated heterogeneous subnetworks

When interconnecting heterogeneous traffic management architectures, two related problems must be solved:

■  The semantics of the traffic contracts of the subnets must match (this has been coined *service model* in [11]). E.g., concerning the quality of service, it is not possible to provide guaranteed service with a subnet that offers only best-effort service, or, concerning the traffic, a bursty traffic source cannot inject traffic in a subnet that assumes constant traffic. With the properly matched concatenated subnets it is possible to formulate the end-to-end traffic contract.

■  When the previous has been solved, a mapping must be found between the parameter sets of the traffic contracts. The user specifies the parameters of the end-to-end traffic contract, these must be mapped on the individual traffic contracts of the subnets. This includes the problem of finding a universal set of parameters that covers the envisaged subnets. Second, this set of parameters must be mappable, more specifically, on both packet-oriented and cell-oriented traffic contracts.

From the last point we derive an important requirement on the signalling protocols: it must carry the parameter set of the end-to-end traffic contract.

## 3.2  Signalling Interworking

### *Signalling Functions*

The first step in trying to find a common approach to resource reservation and negotiation in heterogeneous networks is to study the functionality of existing reservation protocols, namely SCMP and Q.2931. Table 1, "Comparison of signalling protocols," on page 169 provides a comparison of SCMP and the different ATM signalling versions.

Focusing on the functionally richer Forum version of Q.2931 there is a great similarity between Q.2931 and SCMP. Currently SCMP is the richer protocol. In some areas like dynamic QoS changes the function is not available in the current version of Q.2931 but will be added to the next one. More important are the differences in QoS handling schemes. Based on the concept of QoS classes rather than independent QoS parameters in Q.2931, there is no negotiation of QoS values in Q.2931. Also there is no negotiation of traffic parameters in Q.2931, at this point there is only an accept/reject semantic for traffic parameters in ATM. In SCMP all of these parameters are negotiable.

Further SCMP includes intermediate entities like routers as well as end systems into its negotiation scheme, thus if an intermediate router can not support a requested QoS specification, it can reduce the values according to its capabilities. Even more, the SCMP flowspec can be defined to include end- or intermediate system specific parameters like buffer space or the like. Q.2931, targeted at controlling the QoS of a single ATM subnetwork, has no concept of intermediate nodes. User to user negotiation capabilities are very limited due to the fact that user to user parameters in the Q.2931 setup message are limited to 4 bytes, clearly not enough to carry any elaborate QoS parameters set. Knowing that interworking units like routers can add significantly contributions to the end to end transmission delay, the delay variance or to the limitation of throughput it is imperative that intermediate entities as well as end systems are incorporated into the reservation and negotiation scheme. The limitations of Q.2931 in this area are due to the currently defined message formats and could be resolved by appropriately adjusting the format of these messages. Similarly there is no fundamental reason why Q.2931 could not incorporate a true negotiation scheme as opposed to the current accept/reject scheme. Thus today SCMP is the more general signalling protocol, but there is no reason why Q.2931 could not evolve to the same generality.

| feature | ST-II/SCMP | ITU-T Q.2931 Release 1 | ATM-Forum UNI 3.1 Q.2931 Version 1 | future Q.2931 versions |
|---|---|---|---|---|
| call | single connection | single connection | single connection | multiple connections |
| call topology | • bidirectional pt-to-pt <br> • unidirectional pt-to-multipt | • bidirectional pt-to-pt | • bidirectional pt-to-pt <br> • unidirectional pt-to-multipt | • bidirectional pt-to-pt <br> • unidirectional pt-to-multipt <br> • multipt-to-multipt |
| pt-to-multipt setup | • sequential setup <br> • parallel setup | - | • sequential setup <br> • parallel setup | • sequential setup <br> • parallel setup <br> • atomic setup <br> • leaf-initiated joining |
| dynamic calls | • dynamic add/drop of parties <br> • dynamic change of FlowSpec | - | • dynamic add/drop of parties | • dynamic add/drop of parties <br> • dynamic change of traffic contract <br> • dynamic add/remove cnx of a call |
| traffic description | (part of FlowSpec) | • Peak Cell Rate | • Peak Cell Rate <br> • Sustainable Cell Rate <br> • Max. Burst Size | • Peak Cell Rate <br> • Sustainable Cell Rate <br> • Max. Burst Size |
| QoS description | attribute-value pairs (part of FlowSpec) | QoS classes | QoS classes | • QoS classes <br> • attribute-value pairs |
| traffic contract negotiation | • user-network <br> • user-user | - | - | • user-network <br> • user-user |

**Table 1** Comparison of signalling protocols

In addition to the functional requirements discussed above, there are some aspects that are not covered by either SCMP or Q.2931. In particular the ideas of receiver initiated point to multipoint connections and heterogeneous branches in multicast trees as proposed in the RSVP protocol [4] and their requirements on the signalling protocol needs to be investigated further. Both of them will add to the complexity of the signalling protocol but are required to support important application areas like audio/video broadcasting or conferencing.

## Signalling Interworking Architecture Alternatives

When interconnecting heterogeneous networks with a requirement to control resources on a subnet basis, a network layer interconnection architecture is required. Starting from SCMP and Q.2931 base, the following alternatives are possible.

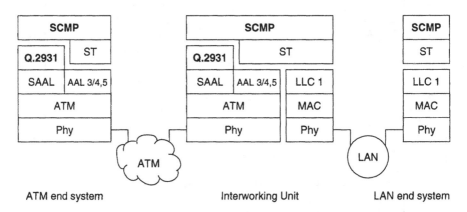

**Figure 4**    ATM as ST-II Subnet

Here the ST-II architecture on the LAN side is unchanged. On the ATM side separate VCCs are used for SCMP messages and ST-II data connections. The message flow for connection establishment is shown in Figure 5. An SCMP connection request arriving at the Interworking Unit triggers the setup of an ATM Virtual Channel Connection (VCC) via Q.2931 signalling. It is followed by a separate SCMP connection establishment over another VCC (not shown in the figure). Obviously this is an extremely inefficient connection establishment scheme, but it does provide the full SCMP functionality in a heterogeneous environment.

The second alternative is the use of ATM like services on the LAN side as shown in Figure 6.

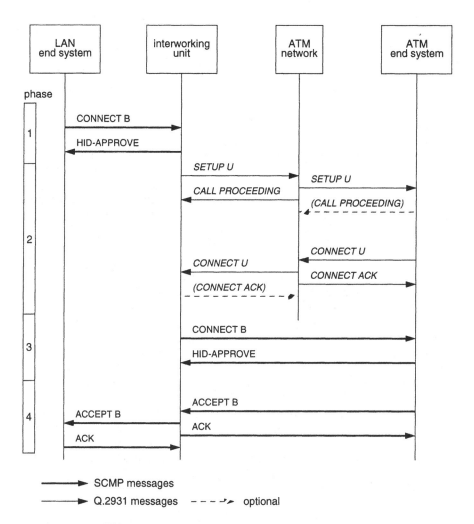

**Figure 5** Connection establishment for the ATM as ST-II Subnet case

phase 1: An ST-II connection request arrives at the Interworking Unit (IWU).
phase 2: The IWU establishes a dedicated ATM Virtual Channel Connection (VCC).
phase 3: The SCMP connection request is forwarded to the next hop.
phase 4: The SCMP connection accept message terminates the connection setup.

In this case a connection oriented LLC 2 can be used to provide a service equivalent to the Signalling AAL on the LAN side. The advantage of this solution is that the layered connection establishment handshakes can be avoided, thus this

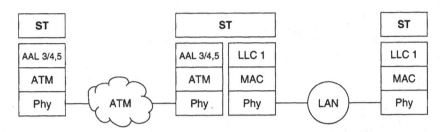

**Figure 6**   Data protocol interworking for the ST-II Subnet case

approach leads to more efficient connection establishment schemes. A second advantage is that a homogeneous signalling protocol will reduce the implementation overhead associated with different schemes. Finally this architecture provides a smooth migration path into the future, when over time less shared media type LANs will be used and eventually a pure ATM networking world may be achieved.

However, it is not a viable alternative with the current state of Q.2931. In particular the above mentioned Q.2931 deficiencies with regard to intermediate and end-system negotiation prohibit this solution. Given the general lack of experience with reservation and negotiation protocols and the lack of a widely accepted standard solution for this purpose, it is unlikely that Q.2931 will evolve over the next few years to provide the generality required to implement this architecture.

The third alternative is a protocol translation approach (see Figure 7).

Again the advantage of this architecture is the more efficient connection establishment scheme, similar as in the previous approach. A second advantage is the lower degree of changes required for existing architectures. Both protocols can be evolved to future versions that need not be identical, but which may be sufficiently similar so that translation between is possible without loss of information.

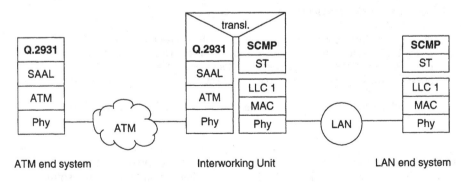

**Figure 7**   Q.2931/SCMP protocol translation

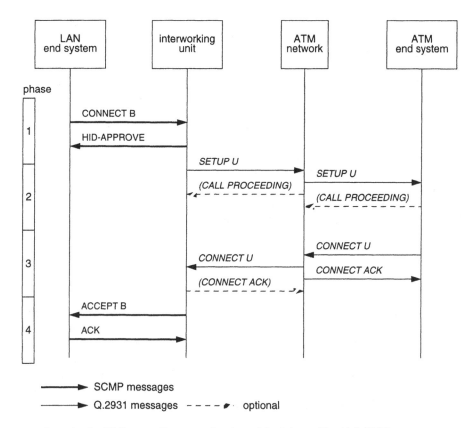

**Figure 8**  Connection establishment
for the Q.2931/SCMP Protocol Translation case

phase 1: An ST-II connection request arrives at the Interworking Unit (IWU).
phase 2: The IWU translates it in an ATM Virtual Channel Connection (VCC) setup request.
phase 3: The request is accepted.
phase 4: The IWU translates the accept message back in an SCMP accept message.

Today this is not the case, for the same reasons that prohibit the previous alternative.

In order to make this architecture implementable, Q.2931 needs to provide more space for user to user QOS parameters. This additional parameters can then be used to carry the information necessary to

- carry a "universally" defined QoS parameter set

- as well as all information necessary to reconstruct the proper SCMP messages at the edge of an ATM network.

In addition the QoS mapping issues described in section 3.1 need to be solved. However, this problem needs to be resolved in any case, independent of the chosen interconnection architecture.

As ATM networks will conquer more and more ground the need for protocol translation will decrease more and more. Thus, from an evolutionary point of view this approach has the same advantage as the previous one. In pure ATM world this architecture leads to an efficient signalling architecture.

Summarizing the above discussion, we conclude that in the current state of SCMP and Q.2931 only a very conventional subnet approach provides the required functionality at the expense of a very inefficient connection establishment scheme. In particular in a pure ATM environment of the future this is not an attractive solution. The protocol translation approach requires a limited set of changes on the Q.2931 side, in order to become viable. Once this is done, it holds the promise for a simpler and more efficient interconnection architecture.

## 3.3  Data Protocol Interworking

The main focus when considering the interworking of heterogeneous networks once all resources have been reserved and all channels have been assigned is on the data transfer efficiency including the interworking unit complexity. This is the main reason to use label swapping based virtual circuit switching techniques for the data phase. Both protocols ST-II and ATM follow this approach. Thus all potentially difficult routing issues need to be resolved during the connection establishment phase. In both protocols there is no flow control on the hop level which could potentially decrease network throughput. A checksum based error control is provided in ATM through the AAL functions and in LANs by the MAC or LLC type 1 layer. Error recovery is always done at a higher layer protocol. Thus ATM AAL 3/4 or AAL 5 and LAN MAC/LLC.1 service are sufficiently similar for a straight forward interconnection architecture. Analogous to choices for interconnecting the signalling architectures we briefly discuss three different interconnection alternatives.

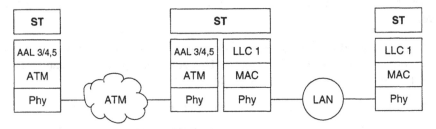

**Figure 9**   Data protocol interworking for the ST-II Subnet case

The first being the ATM as ST-II subnet approach (see Figure 6). In this case the LAN side is unchanged, and on the ATM side the full ST-II packets are transmitted across the ATM network. This implies that the ATM VCC is assigned to an ST-II hop, carrying some redundant information on the ATM side. Given the high speed of ATM networks, this does not seem to be a serious problem.

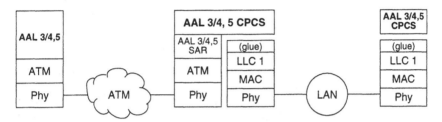

**Figure 10**   Data protocol interworking for the ATM-like Service over LAN case

Certainly there is no point in using ATM cells over LANs, thus the second alternative is to provide the AAL n CPCS (common part of the convergence sublayer) for AAL 5 or AAL 3/4 service over both networks (see Figure 10). This eliminates ST-II messages from the picture, and is only feasible if the same alternative is being used for the signalling part. On the other hand, there is no apparent advantage to use AAL n CPCS instead of ST-II messages on the LAN side. On the ATM side it avoids the need for a separate layer 3 frame, thus it decreases framing overhead on the ATM network.

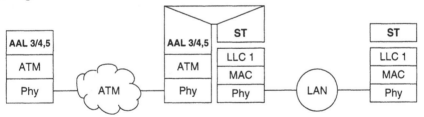

**Figure 11**   Data protocol interworking for the Protocol Translation case

Finally a protocol translation approach can be chosen (see Figure 11). In the absence of layer 3 multiplexing, the mapping between the ATM VC Identifier (VCI) and the ST-II Hop ID (HID) is well defined. A message arriving on a given ATM VCC can be assigned to the correct ST-II connection by its VCI without the need for an explicit HID. The other information contained in the ST-II header field is the message priority and an optional timestamp field. If message priorities have to be supported, the current AAL 3/4 or 5 services are not sufficient. Although AAL 3/4 or 5 provide a cell loss priority bit, this CLP bit is an input parameter for ATM. Once a cell arrives at its destination the CLP is lost and not passed to the receiver. If this was changed, including a consistency check whether all cells of a AAL SDU carry the same CLP, the CLP could be utilized to implement a two level message priority scheme. The optional timestamping field is not recoverable from an incoming ATM message. However, it is questionable how useful network layer timestamping is in any case. Application specific synchronization mechanisms will probably be used in and above an end to end protocol layer.

When comparing the three different data protocol interworking approaches, it appears that the ST-II subnet approach is straight forward and generates only a limited amount of overhead. A protocol translation avoids the layer three framing overhead, thus will be the preferred architecture with regard to the evolution towards pure ATM networks. It does however require a slight modification of the AAL 3/4 or 5 definition, namely the presentation of the CLP for incoming packets.

A last issue, independent of ATM and ST-II shall be pointed out here. Error recovery by packet retransmit is not an acceptable technique for multimedia streams, as it always increases the delay variance (jitter) of the stream. Thus it is currently under discussion whether forward error correction techniques integrated into the AAL are a way to minimize the number of delayed or lost messages. The same arguments will however apply to normal LANs. At this point application requirements or tolerance with regard to lost or corrupted multimedia data need to be studied. The results of these discussions may lead to the definition of a new AAL which is particularly suited for packetized multimedia traffic.

# 4   IMPLEMENTATION OUTLOOK

The problem of interconnecting heterogeneous networks to support multimedia communication is currently being investigated as part of the German BERKOM project. BERKOM (BERliner KOMmunikationssystem) is a multivendor Broadband-ISDN pilot project lead by DeTeBERKOM, a daughter company of the German Telekom. Within the framework of BERKOM a multimedia communication system based on an ST-II network and an XTP derived transport layer is being developed [12]. The first version of this protocol stack is operational on DEC, HP, IBM, Siemens/SNI and SUN Workstations running over Ethernet, FDDI and a synchronous 140 Mbit/s wide area network called VBN. The next step in this project is now being launched in interconnecting Ethernet and FDDI based workstations over local and wide area ATM networks. The target configuration of this trial is shown in Figure 12.

In this configuration the local area networks (Ethernet, Token Ring and FDDI) will be connected via ST-II routers to local ATM switches. An interconnection of

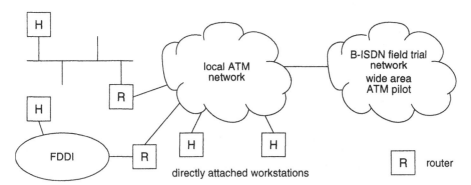

**Figure 12** BERKOM B-ISDN pilot network configuration

local ATM switches over the B-ISDN pilot has already been demonstrated at the CeBIT fair 1994 in Hannover. This connection has been based on preconfigured permanent virtual circuits (PVCs) on the local and on the wide area ATM network. As we do not expect to have signalling support for switched virtual circuits (SVCs) available on the wide area ATM pilot before late 1995, the current plan is to implement the above scenario in the following stages:

1. Ethernet/FDDI and directly ATM LAN attached workstations connected over PVCs

2. same configuration with dynamic connections (SVCs) over the ATM LAN

3. SVCs on the ATM LANs connected through wide area PVCs, in this case separate PVCs will be used for signalling and user data over the wide area ATM network.

4. finally SVCs will be used on local and wide area ATM networks.

Based on the architectural constraints discussed in the previous section, this implementation will first be based on an ATM as ST-II subnet approach, both for the signalling as well as for the data protocol interworking. Based on the experience with the complexity and performance of this implementation and the directions the Q.2931 and SCMP evolution takes over the next years a translational approach can be incorporated at a later stage.

# 5 ACKNOWLEDGMENT

The authors are indebted to Prof. Rainer Oechsle for many helpful discussions and suggestions during the early stages of this work.

# REFERENCES

[1]   C. Topolcic (Editor), "Experimental Internet Stream Protocol, Version 2 (ST-II)", RFC 1190, October 1990.

[2]   R. Händel, M.N. Huber, "Integrated Broadband Networks - An Introduction to ATM-Based Networks", Addison-Wesley, 1991.

[3]   J.-Y. Le Boudec, "The Asynchronous Transfer Mode: A Tutorial", *Computer Networks and ISDN Systems*, Vol. 24, No. 4, pp. 279-309, May 1992.

[4]   L. Zhang, S. Deering, D. Estrin, S. Shenker, D. Zappala, "RSVP: A New Resource ReSerVation Protocol", *IEEE Network*, September 1993, pp.8-18

[5]   ITU-TS, Draft Recommendation Q.2931, ITU-TS SG 11/WG 2 Temporary Document, December 1993.

[6]   ATM Forum, ATM User-Network Interface Specification Version 3.0, ATM Forum, 480 San Antonio Road, Suite 100, Mountain View, CA 94040, USA, September 1993

[7]   D. D. Clark, S. Shenker, L. Zhang, "Supporting Real-Time Applications in an Integrated Services Packet Network: Architecture and Mechanism", *Proc. SIGCOMM'92*, October 1992

[8]   R. Nagarajan, C. Vogt, "Guaranteed-Performance Transport of Multimedia Traffic over the Token Ring", *IBM European Networking Center Technical Report No. 43.9201*, December 1991

[9]   R. Guérin, H. Ahmadi, M. Naghshineh, "Equivalent Capacity and its Application to Bandwidth Allocation in High-Speed Networks", *IEEE JSAC*, Vol. 9, No. 7, September 1991

[10] J. Kurose, "Open Issues and Challenges in Providing Quality of Service Guarantees in High-Speed Networks", *ACM Computer Communication Review*, Vol. 23, No. 1, January 1993

[11] S. Shenker, D. D. Clark, L. Zhang, "A Service Model for an Integrated Services Internet", Internet Draft, October 1993

[12] S. Boecking, S. Damaskos, A. Fartmann, B. Weise, G. Hoelzing, I. Milouscheva, J. Sandvoss, M. Zitterbart, "The BERKOM Multimedia Transport System", *IS&T/SPIE's Symposium on Electronic Imaging: Science and Technology*, February 1994

# 11

# FROM BEST EFFORT TO ENHANCED QoS*

A. Danthine, O. Bonaventure[1]

*Institut d'Electricité Montefiore, B-28, Université de Liège,
B-4000, LIEGE, Belgium
Email: danthine@vm1.ulg.ac.be*

## ABSTRACT

The new communication environment has brought new requirements on the qualities of service for which the present "best effort" semantics is inadequate. The response to this evolution goes through the definition of a new model of QoS for the lower layers. In the "best effort" model, when a service provider accepts a transmission with a given QoS, it does not commit itself to any duty about the way it will take account of this QoS. The "guaranteed" QoS requires resources reservation mechanisms which are not always available. We present in this paper new negotiation rules and a new semantics for the QoS. It allows a service user to express more accurately its requirements and implies for the service provider an obligation of behaviour.

## 1 THE QoS MODEL

The goal of this paper is to study the quality of the service offered by a transport service provider to transport service users, in a peer-to-peer connection mode service. The figure 1 represents the model that will be used in this study. The only observable events are the occurrences of transport service primitives at the two transport service access points (T-SAP).

The term Quality of Service (QoS) is the collective name given to certain characteristics associated with the invocations of the service facilities as observed at the SAP. The QoS is specified in terms of a set of parameters. The performance QoS parameters, in particular, are those QoS parameters whose definition is based on a performance criterion to be assessed.

---

* This work has been done in the framework of the RACE project 2060, CIO.
[1] Research Assistant of the University of Liège

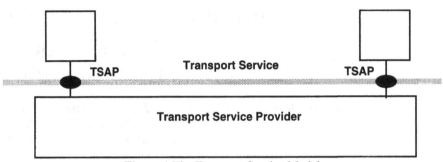

**Figure 1** The Transport Service Model

The role of the QoS becomes increasingly important with the present evolution of the applications. The client/server-based applications demand low-latency request/response-oriented services. The multimedia applications, with their particular needs on throughput and quality of transmission, tend to extent on local, or even wider, networks.

For the service user[2], the quality of service is determined by the values of some end-to-end parameters. The most important performance criteria for a connection are related to the throughput, the transit delay and the reliability.

Of course the assessment of a performance criterion requires the introduction of timing considerations. Since the only events observable by a service user, when it is using a service facility, are the primitives that occur at its SAP, the only notion of time which can be relied on to introduce timing considerations is the notion of time of occurrence of a service primitive at a SAP. Such a time is an absolute time, not usable in isolation.

Time has to appear in the definition of the performance QoS parameters as time intervals between occurrences of service primitives. Thus, any performance QoS parameter has to be defined in relation with the occurrences of two or more related service primitives. They may be related because they pertain to a same connection, or because they are occurring successively at a same SAP, or because one or several parameters of one of the primitives occurring at a given SAP have been replicated in the other primitive(s) occurring at peer SAP(s), etc. The figure 2 gives examples of time intervals usable in relation with the performance parameter definition. More complex relationships resulting from combinations of several basic relationships are also possible.

---

[2] In the following, we will use "service user" and "service provider" instead of "transport service user" and "transport service provider"

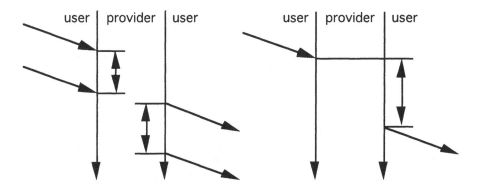

**Figure 2** Time intervals usable in relation with the performance parameter definition

In the connection mode, the values of the QoS parameters are negotiated at the connection establishment time. In the connectionless-mode, the QoS are selected by the calling user and related to a single SDU.

# 2 TYPES OF QoS NEGOTIATIONS

In the peer-to-peer case, the three actors of the negotiation are the calling (service) user, the called (service) user and the service provider. All negotiations are based on the classical 4-primitive exchange; request, indication, response and confirmation .

For some performance parameters such as the throughput, the higher the better. For some other performance parameters such as the transit delay jitter, the smaller the better. Through this text, we will use the terms "weakening" and "strengthening" a performance parameter to indicate the trend of the modification. Weakening a throughput means reducing its value but weakening a transit delay jitter means increasing its value.

## 2.1 Triangular Negotiation for Information Exchange

In this type of negotiation, the calling service user introduces, in the request primitive, the value of a QoS parameter. This value may be considered as a suggested value because the service provider is free to weaken it at will, before presenting the new value to the called user through an indication primitive. The called user may also weaken the value of the parameter before introducing it in the response primitive. This final value will be included without change by the service provider in the confirm primitive. At the end of the negotiation, the three actors have the same value of this QoS parameter (fig. 3).

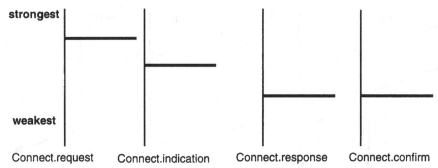

strongest

weakest

Connect.request        Connect.indication        Connect.response        Connect.confirm

**Figure 3** Triangular Negotiation for Information Exchange

Taking account of the freedom for the service provider to weaken the value suggested by the calling user, the service provider will reject directly the request only if it is unable to offer the service whatever the value of the QoS.

The calling user has always the possibility to request a disconnection if it is unsatisfied by the value resulting from the negotiation .

The goal of such triangular negotiation is essentially to exchange information among the three actors and to fix the weakest value agreeable to the three actors.

The ISO Transport Service uses this type of negotiation for the performance-oriented QoS [ISO 8072]. The classes 1 and 3 of the ISO Network Service [ISO 8348] are also based on the same scheme.

## 2.2 Triangular Negotiation for a Bounded Target

In this type of negotiation, the calling user introduces, in the request primitive, two values of a QoS parameter, the target and the lowest quality acceptable. The service provider is not allowed to change the value of the lowest quality acceptable. Here, the service provider is free, as long as it does not weaken it below the lowest quality acceptable, to weaken the target value before presenting to the called user, through an indication primitive, the new value of the target and the unchanged value of the lowest quality acceptable. It will be the privilege of the called user to take the final decision concerning the selected value. This selected value of the QoS will be returned by the called user in the response primitive. This selected value will be included without change by the service provider in the confirm primitive. At the end of the negotiation, the three actors have agreed on the value of this QoS parameter (fig. 4).

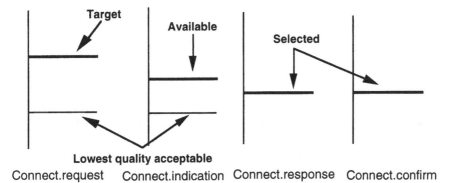

Connect.request   Connect.indication  Connect.response  Connect.confirm

**Figure 4** Triangular Negotiation for a Bounded Target

The service provider may have to reject the request if it does not agree to provide a QoS in the requested range. The called user may also reject the connection attempt if it is not satisfied with the range of values proposed in the indication primitive.

With respect to the target value introduced by the calling user, the only possible modification introduced by the negotiation is the weakening of the target but limited by the lowest quality acceptable value.

The class 2 of the ISO Network Service [ISO 8348] is based on this scheme which is also used in ST-II [Top 90] and [FRV 93].

## 2.3 Triangular Negotiation for a Contractual Value

In this type of negotiation, the goal is to obtain a contractual value of a QoS parameter which will bind both the service provider and the service users. Here the calling user introduces, in the request primitive, two values of a QoS parameter, the minimal requested value and the bound for strengthening it. If it accepts the request, the service provider is not allowed to change the value of the minimal requested value. However the service provider is free, as long as it does not weaken it below the minimal requested value, to weaken the bound for strengthening before presenting to the called user, through an indication primitive, the new value of the bound for strengthening and the unchanged value of the minimal requested value (fig 5). It will be the privilege of the called user to take the final decision concerning the selected value. This selected value of the QoS will be returned by the called user in the response primitive and is acceptable for the service provider. This selected value will be included, without change by the service provider, in the confirm primitive. At the end of the negotiation, the three actors have agreed on the value of this QoS parameter (fig 5).

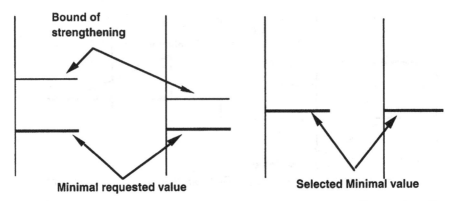

Figure 5 Triangular Negotiation for a Contractual value

The service provider may have to reject the request if it does not agree to provide a QoS in the requested range. The called user may also reject the connection attempt if it is not satisfied with the range of values proposed in the indication primitive.

With respect to the minimal requested value introduced by the calling user, the only possible modification introduced by the negotiation is the strengthening of the minimal requested value but limited by the bound for strengthening value. The service provider may weaken the bound for strengthening and the called user may strengthen the minimal requested value but up to the limit accepted by the service provider.

It is this scheme of negotiation that is used in OSI95 for two types of requested values.

## 2.4 Bilateral Negotiation

Figure 6   Bilateral Negotiation

For some QoS parameters, the negotiation takes place between the two service users, the service provider being not allowed to modify the value proposed by the service user (figure 6). However, this bilateral character is more formal than real, the service provider having always the possibility to reject the request.

## 2.5 Unilateral Negotiation

When not only the service provider but also the called service user are not allowed to modify the value proposed by the calling user, the negotiation is reduced to a "take it or leave it" approach (fig 7). This type of negotiation may have its merits in special situations.

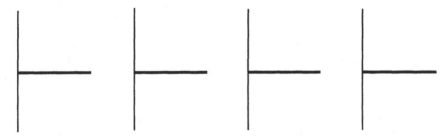

Connect.request  Connect.indication  Connect.response  Connect.confirm

**Figure 7**  Unilateral Negotiation

## 3 BEST EFFORT QoS

## 3.1 The Semantics of the QoS in the ISO Transport Service

Coming back to the result of the negotiation for information exchange of § 2.1, it is clear that it is difficult to attach a strong semantics to the resulting value of the QoS parameter. It is only a value on which all three actors agreed.

The semantics of the QoS in the ISO Transport Service is that of the Best Effort, and *"users of the Transport Service should be aware that there is no guarantee that the originally negotiated QoS will be maintained throughout the Transport Connection lifetime, and that changes in QoS are not explicitly signalled by the Transport Service provider."* [ISO 8072]

Even when the calling TS user prohibits the negotiation of a particular QoS parameter and therefore, expresses its QoS as an "absolute" requirement in the CONNECT.request, the service provider will still be allowed to violate the QoS value without notice during the lifetime of the connection.

The QoS parameter does not require a permanent monitoring by the service provider because it is not possible to specify a particular behaviour of the service provider if the real value of the QoS parameter is weaker than the agreed value. The service users do not expect any particular behaviour in such case. They are just expecting the *"best effort"* from the service provider.

In such a loosely defined environment, if a service user introduces, in a request primitive, the value of a QoS parameter, it is not always clear whether this suggested value is related to a boundary or to an average value. In the latter case, the measurement sample or the number of SDUs to be considered is often far from obvious. However, this is not a great problem if no monitoring has to be done.

For a negotiation in which the service provider is not allowed to weaken the value suggested by the calling user, it is possible for the service provider to reject the request due to the requested QoS value but, as already mentioned, in case of acceptance, nothing will be done if the service provider does not reach the QoS value. In this case, the service user is not even informed about the situation by the service provider as no REPORT indication primitive has been defined.

It is therefore not surprising that in many cases, the QoS is expressed in qualitative terms without any specification of a given value. This confirms the lack of relationship between the QoS parameter and a real performance parameter. The only way for the service users to assess the value of a QoS is to monitor it.

The situation we just described is the today situation of transport service and transport protocols in ISO.

If, to operate in a correct way, an application requires a well-defined set of performance parameters, the present approach will not be suitable.

## 3.2 The Monitoring and the Protocol Functions

With such a poor semantics of the performance QoS parameters of the ISO TS, it may be surprising that the situation has been accepted for more than 10 years.

This results from the fact that some performance QoS values result directly from the protocol functions which are implemented by the service provider. The best example of this situation is the error control scheme in protocols such as TP4 [ISO 8073]. The goal of this protocol is to deliver the data in order and without corruption to the receiving user. To achieve this, TP4 uses two protocol functions : the detection of errors and the retransmission of the lost or corrupted data. The errors are detected with a checksum that covers each DATA TPDU.

This checksum can only detect errors with a known probability which depends on the length of the DATA TPDUs covered by the checksum. When a DATA TPDU has been lost or corrupted, it will have to be retransmitted. With these protocol functions with known (i.e. mathematically provable) properties, TP4 achieves its goal with a very low value of the residual error rate. When the protocol detects (usually after a certain amount of retransmissions without acknowledgement) that it is no more able to transmit the data reliably, it releases the connection.

In such a situation, the monitoring of the residual error rate by the service provider is impossible and the associated QoS negotiation is without interest because requiring a residual error rate weaker than the value provided by the error control function is without interest and requiring a residual error rate stronger than the value provided by the error control functions may not even force the service provider to reject the connection request.

For the data communication, the ISO transport service has been considered as adequate because it was offering a reliable service which was the basic requirement. The throughput, the transit delay and the transit delay jitter were not considered as critical factors of performance and the "best effort" situation we just described was acceptable for most of the past applications.

# 4 THE NEED FOR AN ENHANCEMENT

In [Dan 94], the need for new standards or at least for an enhancement of existing ones, has been analysed, based on the consequences of the changes we are facing in network performance, in network services and in the application environment.

At the service level, the need for an enhancement means the need for a new semantics for the QoS. If, in the past, the basic requirement, at the transport service level, was a reliable transfer of data associated with a best effort for the other performance QoS parameters, the situation has to be improved.

A new application based on a client-server paradigm may have a specific requirement for the transit delay and the round-trip time.

Some multimedia-based applications will not be able to operate properly if the throughput offered by the transport service falls below some specific value.

The time dependence between successive video frames may not be preserved if the delay jitter goes above a given value.

Those three examples are mentioned to show that the best effort approach for the throughput, the transit delay and the transit delay jitter may not be an acceptable mode of operation for the service provider.

To enhance the present situation, it is possible to associate with the QoS the concept of a guarantee. By so doing we will associate a very strong semantics with the QoS. This has of course some implications which will be analysed in the next section.

## 5 THE GUARANTEED QoS

It would be nice to be able to have the concept of a guaranteed QoS, especially when the guaranteed values result from a negotiation for a contractual value presented in section 2.3.

The possibility for the service provider to give such a guarantee is related to the existence of enough resources associated with some protocol function for allocating the resources and managing them during the connection.

As already discussed, a residual error rate on a connection may be guaranteed by the service provider if enough storage resources can be allocated to the connection and if it operates an error control function with known properties.

A minimum throughput on a connection may be guaranteed by a transport service provider if each transport protocol entity can be allocated enough processing and storage resources and if the underlying network service provides the guarantee of a throughput compatible with the requested one, taking into account the layer overhead (fig. 8).

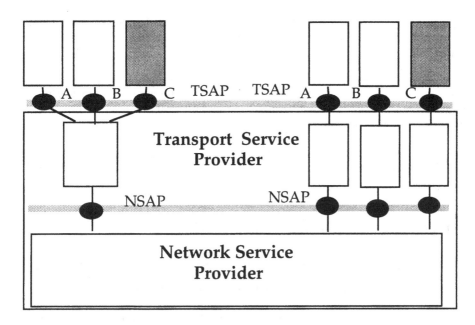

**Figure 8** The Transport Service and the Network Service

With the best effort, a request to establish the transport connection CC (fig. 8), will usually be handled without taking account of the resources already used by the AA and BB connections and of the limited throughput of the network service provider. The result will be that the three connections will share the processing resources of the transport entities and will also share the throughput capacity of the network service provider. Any opening of a new connection will increase the level of multiplexing at the transport level and may affect the performance QoS parameters of the already established connections.

With a guaranteed QoS semantics, the situation will be different. Any attempt to establish a new connection with a guaranteed throughput (such as CC in figure 8) will imply the evaluation of the remaining processing and storage capacities of the transport entities and of the possibility for the network service provider to guarantee the needed throughput taking account of the layer overhead.

The basic point here is the permanent availability of resources allocated to the connection. It is only in this case that the service provider will be able to "guarantee" the QoS value requested by the service users. In this case, the monitoring of the QoS values is not necessary as the requested value will be achieved by the service provider, except in exceptional pathological situations.

With a guaranteed QoS semantics, the transport service, if not able to allocate the necessary resources, may be obliged to reject a request to establish the transport connection CC in order to protect the QoS of the already established connections. The same non pre-emptive approach has also been proposed in [FRV 93].

In practice, it is not obvious to obtain a guaranteed throughput from the network service even when the network service is provided by a single subnet. This is however possible with the synchronous service of FDDI and the ATM may also offer an equivalent service.

When the network service is provided by an internet, it is almost impossible to get today any guarantee on a throughput. This is the case with IP and with the ISO internet.

## 5.1 The Internet Stream Protocol, ST-II

As IP offers a pure connectionless service, it cannot provide any guarantee on the achieved QoS nor allow resources to be reserved. TCP only adds reliability above the IP layer, but it does not support other QoS parameters.

Proposals for an extended IP protocol [Dee93], [Fra93] recognise the need to include Quality of Service parameters in an extended IP, but they do not completely support it now because there is no common understanding yet of what kind of QoS is needed in such an extended IP. If this extended IP remains entirely connectionless, there is little hope that it will offer any guarantee on the QoS. [Dee 93] states that a new reservation protocol is being developed for SIP. This protocol will be used to set-up a route and reserve resources on each node along the route.

This was already the philosophy of [AHS 90] and of the Internet Stream Protocol, Version 2 (ST-II) which is "*an IP-layer protocol that provides end-to-end guaranteed service across an internet.*" [Top 90]. One of the main goals of ST-II was to provide a point-to-point simplex and a point-to-multipoint simplex data transfer for the applications with real-time requirements. In the ST-II specification, these simplex data paths are called "stream".

*"An ST stream is :*

*- the set of paths that data generated by an application entity traverses on its way to its peer application entity(s) that receive it,*

*- the resources allocated to support that transmission of data, and*

*- the state information that is maintained describing that transmission of data."*

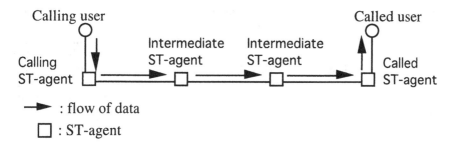

**Figure 9** Point-to-Point Simplex ST-II stream

In ST-II, the first step is to set a route along which resources are allocated and to activate ST-agents (fig 9).

ST-II uses a data structure named flow specification (abbreviated as flowspec). This flowspec is used to carry the target and the minimal values of the QoS parameters requested by the calling user. It is also used to carry values introduced by the service provider. The flowspec is passed from ST-agent to ST-agent as the path is defined (fig 9). Every ST-agent along the path has to reserve enough resources based on the requested values, to adjust the target values of the negotiation scheme of §2.2 and to estimate its contribution to cumulative QoS parameters such as minimum, mean and maximum transit delay. The global information about the path contained in the flowspec is presented to the called user in the CONNECT.indication. The called user will then have to decide whether the performance values are sufficient before accepting the connection.

The problem with ST-II is related to the type of reservation : firm or pre-emptive, statistical or incomplete [CSZ 92], [Par 92]. In most cases, it is not possible to attach a clear semantics to the QoS due to the lack of firm reservation.

## 5.2 What if?

If the semantics of best effort is associated with the QoS value resulting from a negotiation for information exchange, the service provider has no obligation and nothing will be done if the agreed value is not achieved. The service provider may not even be aware of the situation as it is not obliged to monitor the achieved QoS values.

If the semantics of guaranteed QoS is requested by the service user, the service provider has to reject the request if it does not agree to provide a QoS in the requested range.

Once the request has been accepted, as the result of a negotiation for a bounded target or of a negotiation for a contractual value, the service provider will guarantee the negotiated QoS value and, unless (unavoidable and unpredictable) pathological cases, will provide the requested service. Here, **the service provider has an obligation of results.** This should be the only meaning of a guaranteed QoS and with this strong semantics, there is no need for monitoring, the service provider being certain, at the end of the negotiation, of its ability to provide the requested value[3].

# 6 THE QoS ENHANCEMENTS IN OSI95

If a guaranteed QoS is not possible to achieve, it does not mean that the only choice left is the best effort. There is a room for an enhanced QoS semantics. If the service provider is not able, for any reason, to guarantee the requested QoS, it may try to provide the requested QoS value but will have **an obligation of behaviour** if it does not succeed. In this case, the service provider will have to monitor the achieved QoS value.

When the service provider notices that the achieved QoS value is weaker than the negotiated one, it will have to take some action, such as aborting the connection if it has been instructed to do so by the service user.

The service provider may have been instructed not to abort, but to indicate to the service user(s) that it cannot maintain the selected value, and leave to the service user(s) the responsibility of aborting.

It is on this basis that, in OSI95, a new semantics for the performance QoS parameters has been introduced. In this semantics, a parameter is seen as a structure of three values, respectively called "compulsory", "threshold" and "maximal quality", all these values being optional. Each one has its own well-defined meaning, and expresses a contract between the service users and the service provider [DBL 93].

This means, and this is the most fundamental difference with the best effort, that the service provider is now subject to some well-defined duties, known by each side. In other words, the rules of the game are clear.

The existence of a contract between the service users and the service provider implies that, in some cases, the service users are also subject to well-defined duties also derived from the application environment.

---

[3]Without the obligation of results and without the monitoring, we are back to the best effort.

Depending upon the type of facilities, these QoS parameters will be subject to a negotiation or not. In case of a negotiation, some rules related to the rights of strengthening or lowering (weakening) the QoS values are defined for each type of value and for each participant in the negotiation. This has to be done to keep the final result consistent with the meaning of the parameter value and is part of the needed admission control of a new connection, admission control which will be based on the QoS on one side and on the source characterisation on the other side. This connection admission control will have to be completed by an admission control on a connection to enforce the contract.

# 7 THE COMPULSORY QoS VALUE

The idea behind the introduction of a "compulsory" QoS value is the following one: *when a compulsory value has been selected for a QoS parameter of a service facility, the service provider will monitor this parameter and abort the service facility when it notices that it cannot achieve the requested service.*

It must be clear that no obligation of results is linked to the idea of compulsory value. The service provider **tries** to respond to the requested service facility and, by monitoring its execution, it will

- either execute it completely without violating the selected compulsory value of the performance parameter;

- or abort it if the selected compulsory value of the performance parameter is not fulfilled.

This is, by the way, the semantics associated with the error control in TP4 [ISO 8073]. The connection is released after a number of unsuccessful transmission attempts.

## 7.1 Compulsory QoS versus Guaranteed QoS

The guaranteed QoS has a stronger semantics. When a guaranteed QoS value has been selected for a parameter of a service facility, the service provider will execute completely the service facility without violating the selected guaranteed value of the performance parameter.

The compulsory concept reflects the fact that, in some environments (e.g. a lightly loaded LAN), the compulsory QoS value may be achieved without resource reservation. Of course, the same LAN, which does not provide any reservation mechanism or any priority mechanism, may, when heavily loaded, prevent the service provider from reaching the compulsory QoS value and oblige it to abort the execution of the requested service facility.

## 7.2 QoS Parameters and Information Parameters

The introduction of compulsory QoS values implies that the service provider will have a more difficult task to fulfil. It is therefore not surprising that the service user may have to provide the service provider with more information about the characteristics of the elements associated with the request in order to facilitate the decision of rejection or acceptance of the request. Requesting a throughput of 2 Mb/s with SDUs of 10 KBytes is different from requesting a throughput of 2 Mb/s with SDUs of 40 bytes.

Hence, the introduction of the concept of compulsory QoS requires the introduction, in the primitives associated with a request, of additional parameters. These additional parameters may be designated as information parameters to distinguish them from the QoS parameters proper. Information parameters will be used for instance for source characterisation.

Values of QoS information parameters may also have to be introduced to control the negotiation process to preserve the semantics associated with the negotiated value.

## 7.3 Negotiation of a Compulsory QoS value

The negotiation for a compulsory QoS value will follow the negotiation scheme for a contractual value introduced in § 2.3 When a service user introduces a compulsory QoS value for a performance parameter to be negotiated, the only possible modification is the strengthening of this compulsory value. In particular, it is absolutely excluded for the service provider to modify this value in order to relax the requirement

As the calling service user may not be interested in an unlimited strengthening of the proposed compulsory QoS value. It introduces therefore in the request primitive, a second value which fixes a bound indicating to what extent the proposed compulsory QoS value may be strengthened (fig. 5).

When the service provider analyses the request of the calling service user, it has to decide whether it rejects it or not (it can already do so as it knows that the request could only be strengthened). In the latter case, it has to examine the bound of strengthening. This bound may be made poorer (brought closer to the compulsory value) by the service provider, before issuing the indication primitive to the called service user, in such a way to give, to the called service user, the range of compulsory values acceptable by both the calling service user **and** the service provider.

The service provider does not have to strengthen the compulsory QoS value which must be seen as the expression of the requirements of the service users.

After receiving the indication primitive, the called service user may accept or reject the request. If it accepts it, it may modify (strengthen) the compulsory QoS value up to the value of the bound and return it in its response. In this case the negotiation is completed and the service provider may confirm the acceptance of the request and provide the final selected compulsory QoS value to the calling service user.

If the negotiation is successful, the bound is of no interest anymore (the bound is an example of information values mentioned earlier) and the selected compulsory QoS value reflects now the final and global request to the service provider from both service users.

# 8 THE THRESHOLD QoS VALUE

Some service users may find that the solution of aborting the requested service facility, when one of the compulsory QoS values is not reached, is a little too radical. They may prefer to get information about the degradation of the QoS value.

To achieve that we introduced a "threshold" QoS value with the following semantics: *when a threshold value has been selected for a QoS parameter of a service facility, the service provider will monitor this parameter and indicate to the service user(s) when it notices that it cannot achieve the selected value.*

This threshold QoS value may be used without an associated compulsory value. In this case, the behaviour of the service provider is very similar to the one it has to adopt with a compulsory value. The main difference is that, instead of aborting the service facility when it notices it is unable to provide the specified value, it warns either or both users depending of the service definition. If the service provider is able to provide a QoS value better than the threshold value, everything is fine.

## 8.1 Threshold QoS versus Best Effort QoS

If the threshold QoS is used without any compulsory QoS, the main difference between the threshold and the best effort is that in the former case, the service provider has the obligation to monitor the parameter and to indicate if the threshold value is not reached.

## 8.2 Threshold and Compulsory QoS values

It is possible to associate, with the same QoS parameter, a threshold and a compulsory QoS values with, of course, the threshold value "stronger" than the compulsory one.

## 8.3 Negotiation of a Threshold QoS value

The negotiation procedure of a threshold value is similar to the negotiation procedure of a compulsory value. Here also the only possible modification is the strengthening of the threshold value. Here also the calling service user introduces, in the request primitive, an information parameter which fixes a bound indicating to what extent the proposed threshold QoS value may be strengthened.

If a compulsory and a threshold value are associated with the same QoS parameter, there exists a set of order relationship between the compulsory, the threshold and their bounds values which must be verified in the request primitive and maintained during the negotiation.

## 9 THE MAXIMAL QUALITY QoS VALUE

In most cases, if the service provider is able to offer a "stronger" value of the QoS parameter than the threshold, the service user will not complain about it. But it could happen that the service user wants to put a limit to a "richer" service facility.

A called entity, for instance, may want to put a limit to the data arrival rate or a calling entity may want, for cost reasons, to prevent the use of too many resources by the service provider.

Such a parameter may be useful to smooth the behaviour of the service provider. Introducing a maximal quality QoS value on a transit delay, i.e. fixing a lower bound to the transit delay values will reduce the transit delay jitter and facilitate the resynchronization at the receiving side.

To achieve that we introduced a "maximal quality" QoS value with the following semantics: *when a maximal quality value has been selected for a QoS parameter of a service facility, the service provider will monitor this parameter and avoid occurrence of interactions with the service users that would give rise to a violation of the selected value.*

It is possible to associate, with the same QoS parameter, a maximal quality, a threshold and a compulsory QoS values with, of course, the maximal quality "stronger" than the threshold value, itself "stronger" than the compulsory value.

## 9.1 Negotiation of a Maximal Quality QoS value

If a service user introduces a maximal quality QoS value for a performance parameter, the only possible modification is the weakening of this maximal quality QoS value. This value can be weakened during the negotiation by the service provider that indicates by this way the limit of the service it may provide and by the called service user.

If the maximal quality is the only value associated with a given QoS parameter, no bound will be introduced in the request primitive and the negotiation will result in the selection of the weakest of the maximal quality values of the service users and the service provider following the negotiation scheme of § 2.1.

If the maximal quality value and a compulsory or/and a threshold values are associated with the same QoS parameter, there exists an order relationship between the maximal quality value and the bound value on the threshold (or the bound value on the compulsory value if no threshold value is specified) which must be verified in the request primitive and preserved during the negotiation.

## 10 THE OSI95 TRANSPORT SERVICE

In the OSI95 Transport Service [DBL 92c], a QoS parameter is seen as a structure of the compulsory, threshold and maximal quality values, all these values being optional. The main QoS parameters of the OSI95 Transport Service are the throughput, the transit delay, the transit delay jitter and the error control. These QoS parameters have been precisely defined. In [DaB 93], the QoS parameters of the OSI95 Transport Service are compared to other proposals for the QoS parameters in the network and transport services.

The OSI95 connection-mode transport service [BBL 94] formally specified in LOTOS [BLL 92b] has been the starting point for the ongoing development of two enhanced transport protocols.

The first one is assuming a connectionless network service of the IP or CLNS type. This transport protocol implements the negotiation schema discussed earlier and takes into account :
 - the storage and computing resources of each transport entities
 - the maximum bound of the access rate of the network service below each transport entity

- the QoS values achieved in the past on the connections between the two service users
- the characteristics of the QoS associated with the ongoing connections. In particular, the connection request will be rejected if its establishment is likely to jeopardise the compulsory QoS values of already established connections.

This transport protocol monitors the non-(best effort) QoS parameters.

The flow control function based on windows and positive acknowledgements is operating very often in a stop-and-go fashion, not favourable to the multimedia based applications. Like other works on high speed transport protocols, we favour rate control as the basic function without rejecting the possibility to integrate flow control to iron out the allowed differences of behaviour of the two ends of the transport connection.

The second transport protocol under study assumes an underlying service based on ATM switched virtual circuits. Here, instead of having to assume a value for the global access rate at the service access points, it is possible to map the QoS request at the transport service in the request of guaranteed QoS of an ATM virtual circuit.

This second approach intends to use the ATM service to provide an enhanced transport service and is opposite to the approach of the ATM Forum with the "ATM LAN emulation" and of the IETF with "IP over ATM" which are trying to hide the ATM QoS to preserve the old service definitions

# 11 CONCLUSION

After having introduced the QoS paradigm, the principle of the QoS parameter definition and a new taxonomy to distinguish different types of negotiation of the QoS, we discussed the limitation of the best effort QoS and the resources reservation constraints associated with the guaranteed QoS.

We presented in this paper an enhancement of the QoS semantics able to match the communication requirements of the new application environment. This enhanced semantics is based on compulsory, threshold and maximum values and involves a new negotiation scheme aiming at the definition of the contract between the service user and the service provider. The table 1 below summarises the behaviour of the service provider with the various semantics discussed.

|  | Best Effort | Maximum Quality | Threshold | Compulsory | Guaranteed |
|---|---|---|---|---|---|
| Obligation of Results | NO | YES | NO | NO | YES |
| Monitoring | NO | YES | YES | YES | NO |
| Disconnect if achieved value is weaker | NO | N/A | NO | YES | N/A |
| Indicate if achieved value is weaker | NO | N/A | YES | NO | N/A |

N/A : Not applicable

**Table 1** Behaviour of the service provider with the different QoS semantics

This work is at the origin of the OSI95 Transport Service [Dan 92a, 92b], [DBL 92a, 92b], [BLL 92a, 92b, 94] and of the corresponding transport protocols currently under development.

This enhancement of the QoS semantics has also been presented to the standardisation committees [DBL 93], [ISO 6/8010] and we hope that it will contribute to the specification of new communication services.

This work on QoS is presently being extended to cover the multicast case.

# REFERENCES

[AHS 90] D.P. Anderson, R.G. Herrtwich, C. Schaefer, "SRP : A Resource Reservation Protocol for Guaranteed-Performance Communication in the Internet", TR-90-006 International Computer Science Institute, February 1990, 26p.

[ATV 91] M.E. Anagnostou, M.E. Theologou, K.M. Vlakos, "Quality of service requirements in ATM-based B-ISDNs", Computer Comm., Vol. 14, N° 4, May 1991, pp. 197-204

[BLL 92a]   Y. Baguette, L. Leonard, G. Leduc, A. Danthine, "Belgian National Body Contribution - Four Types of Enhanced Transport Services and their LOTOS Specifications", *SC6 plenary meeting, San Diego, July 8-22, 1992*, ISO/IEC JTC1/SC6 N7323, May 1992, 58 p. (OSI95/ULg/A/22/TR/P, May 1992, SART 92/12/09).

[BLL 92b]   Y. Baguette, L. Leonard, G. Leduc, A. Danthine, O. Bonaventure, "OSI95 Enhanced Transport Facilities and Functions", OSI95/Deliverable ULg-A/P, Dec.1992, 277 p. (SART 92/25/05)

[BLL 94]    Y. Baguette, L. Leonard, G. Leduc, A. Danthine, The OSI95 Connection-Mode Transport Service, *The OSI95 Transport Service with Multimedia Support*, A. Danthine, ed., Springer Verlag, pp 181-198.

[CSZ 92]    D.D. Clark, S. Shenker, L. Zhang, "Supporting Real-Time Applications in an Integrated Services Packet Network : Architecture and Mechanism", SIGCOMM'92, Communications Architectures and Protocols, Baltimore, August 17-20, 1992, pp. 14-24

[DaB 93]    A. Danthine, O. Bonaventure, "From Best Effort to Enhanced QoS", Deliverable R2060/ULg/CIO/DS/P/004/b1 of the RACE CIO project, 51 p. (SART 93/15/15) - also circulated as ISO/IEC JTC1/SC6/WG4 N827

[Dan 92a]   A. Danthine, "A New Transport Protocol for the Broadband Environment", *IFIP Workshop on Broadband Communications*, Estoril, January 20-22, 1992, A. Casaca, ed., Elsevier (North-Holland), 1992, pp. 337-360.

[Dan 92b]   A. Danthine, "Esprit Project OSI95 - New Transport Services for High-Speed Networking", *3rd Joint European Networking Conference*, Innsbruck, May 11-14 1992, also in: *Computer Networks and ISDN Systems 25 (4-5)*, Elsevier Science Publishers (North-Holland), Amsterdam, November 1992, pp. 384-399.

[Dan 94]    A. Danthine, "The Networking Environment of the Nineties and the Need of New Standards", *The OSI95 Transport Service with Multimedia Support*, A. Danthine, ed., Springer Verlag, pp 1-12.

[DBL 92a]   A. Danthine, Y. Baguette, G. Leduc, "Belgian National Body Contribution - Issues Surrounding the Specification of High-Speed Transport Service and Protocol", *SC6 interim meeting, Paris, February 10-13, 1992*, ISO/IEC JTC1/SC6 N7312, January 1992 58 p.(OSI95/ULg/A/15/TR/P/V2, 46 p., January 1992, SART SART 92/04/05).

[DBL 92b]   A. Danthine, Y. Baguette, G. Leduc, L. Leonard, "Belgian National Body Contribution - The Enhanced Connection-Mode Transport Service of OSI95", *SC6 plenary meeting, San Diego, July 8-22, 1992*, ISO/IEC JTC1/SC6 N7759, June 1992  16 p. (OSI95, OSI95/ULg/A/24/TR/P, 16 p., June 1992, SART 92/14/05).

[DBL 92c] A. Danthine, Y. Baguette, G. Leduc, L. Leonard, "The OSI95 Connection-mode Transport Service - The Enhanced QoS", *IFIP Conf. on High Performance Networking*, Liège, December 16-18, 1992, in: A. Danthine, O. Spaniol, eds., C14 High Performance Networking, Elsevier Science Publ. (North-Holland), Amsterdam, 1993, pp. 235-252.

[DBL 93] A. Danthine, Y. Baguette, L. Leonard, G. Leduc, "An Enhancement of the QoS Concept", Jan. 1993, 16 p. (OSI95/ULg/A/28/TR/P) - see also ISO/IEC JTC1/SC6/N8010

[Dee 93] S. Deering, "SIP: Simple Internet Protocol", IEEE Network Magazine, Vol. 7 No. 3 (May 1993)

[Fer 92] D. Ferrari, "Real-Time Communication in an Internetwork", Journal of High Speed Networks, Vol. 1, N° 1, 1992 pp. 79-103

[Fra 93] P. Francis, "A Near-Term Architecture for Deploying Pip", IEEE Network Magazine, Vol. 7 No. 3 (May 1993)

[ISO 6/8010] "Liaison Statement to ISO/IEC JTC 1/SC 21 on Qualities of Service (QoS) Work", ISO/IEC JTC1/SC6 N8010, 17 March 1993, 16 p.

[ISO 8072] ISO/IEC JTC1, "Transport service definition for Open Systems Interconnection", ISO/IEC JTC1/SC6 N7734 (ISO DIS 8072 - DIS ballot terminates on April 1993).

[ISO 8073] International Standard ISO/IEC 8073, "Connection oriented transport protocol specification" [Third edition , 1992-12-15]

[ISO 8348] ISO/IEC JTC1, "Network Service Definition for Open Systems Interconnection", ISO/IEC JTC1/SC6 N7558, 21 Sep. 1992

[Kes 92] S. Keshav, "Report on the Workshop on Quality of Service Issues in High Speed Networks", Computer Communication Review, Vol. 22, N° 5, October 1992, pp. 74-85

[Kur 93] J. Kurose, "Open Issues and Challenges in Providing Quality of Service Guarantees in High-Speed Networks", ACM Computer Communication Review, Vol. 23, N° 1, January 1993, pp. 6-15

[LaP 91] A.A. Lazar, G. Pacifici, "Control of Resources in Broadband Networks with Quality of Service Guarantees", IEEE Communication Magazine, Vol. 29, N° 10, October 1991, pp. 66-73

[LiG 90b] T.D.C. Little, A. Ghafoor, "Network Considerations for Distributed Multimedia Object Composition and Communication", IEEE Network Magazine, Vol. 4, N° 6, November 1990, pp. 32-49

[Par 92] C. Partridge, "A proposed Flow Specification", Network Working Group RFC 1363, Sep. 1992

[Top 90] C. Topolcic, Ed., "Experimental Internet Stream Protocol, Version 2 (ST-II)", Network Working Group RFC 1190, Oct. 1990

# 12

# END-SYSTEM QOS MANAGEMENT OF MULTIMEDIA APPLICATIONS

W. Tawbi, A. Fladenmuller, E. Horlait

*Laboratoire MASI, Université Pierre et Marie Curie*
*4, place Jussieu, 75252 Paris Cedex 05, France*

## ABSTRACT

This paper presents a quality of service (QoS) management framework dealing with multimedia applications requirements at the end-systems. The framework is based on the AQOSM (Applications QoS Manager) and a QoS negotiation protocol called HLQNP (High Level QoS Negotiation Protocol). The AQOSM and the protocol provide the applications with the means for expressing and dealing with their QoS in a suitable manner and ensure the link with the available resources.

## 1 INTRODUCTION

Distributed multimedia applications handling different kinds of media with varying characteristics impose new requirements on communication systems. These requirements are related to the applications architecture and to the nature of the media they manipulate. Dealing with these needs has led to the development of several protocols and frameworks based on the quality of service (QoS). Quality of service is then considered at different levels beginning from the end-users requirements to the support by the network nodes responsible of multimedia information transmission.

Most of the current works are focused on internal network support of multimedia transmission and on communication protocol issues. There is a lack concerning the management of applications QoS at end-systems. Applications QoS refers to the quality of service issues that are related to the high level application requirements like the quality of a video stream for a given application and the possible degradation of this stream in case the underlying system is not able to support its transmission.

In this paper we study applications QoS issues and we propose a general framework for their support at end-systems. We first identify the important features related to QoS. We provide means for their expression by the application and a model for their management which provides the interaction with end-system resources. The management entity is called the AQOSM (Applications QoS Manager). A protocol is specified for the distributed QoS management of the applications as a part of the AQOSM.

This paper is structured as follows. Section 2 provides an overview of works related to QoS in multimedia communication systems. In section 3, we describe the framework for applications QoS support. Section 4 presents the QoS management protocol used within the framework and section 5 draws out some conclusions.


# 2 QoS IN DISTRIBUTED MULTIMEDIA SYSTEMS

Several works related to quality of service in distributed multimedia systems have been developed recently. In this section, we present a quick overview of these works.

The ITU-T (formerly CCITT) has defined a number of QoS parameters characterizing network services. Some measurements on these parameters, like the opening and closing connection delays are provided [1]. The QoS issues are limited to the network boundary and do not encompass it to application specific considerations. In the OSI reference model, a set of QoS parameters is defined for each layer, but there are no possibilities to guarantee these parameters by the protocols and the services. Other problems are related to QoS mapping between the layers and the semantics of the QoS parameters.

Recently, new QoS extensions have been proposed for the OSI connection-oriented transport service [2]. These extensions concern mainly the new QoS semantics (compulsory and threshold QoS) and new QoS interpretation. The jitter parameter (delay variance) has been added to the set of parameters that can be negotiated by the service. Renegotiation of the QoS of an existing connection without the need of stopping it is also an added feature to the service.

At the network level, several QoS reservation protocols provide QoS guarantees for multimedia information transmission in Internet networks. These protocols, like ST-II [3], RSVP [4] and RCAP [5], are based on the same principle: they allow the establishment of a "channel" for which resources (mainly CPU and network bandwidth) are reserved. Data transmission follows always the same channel (route) and hence can be conveyed with guarantees on QoS parameters specified by the client and negotiated at the establishment phase.

A proposal for an architecture supporting multimedia communications is given in [6]. Four levels are defined and quality of service and multicast issues are considered at each level. The architecture provides functions for QoS negotiation, monitoring and policing. The first level, the application platform and operating system, is an ODP (Open Distributed Processing) compatible platform that allows an abstract definition of QoS in application related terms. The next level, the orchestration level, deals with multimedia synchronization problems. The transport sub-system level offers, among others, a continuous media service for multimedia information transfer. Finally, the network level offers a multiplicity of network services adapted to higher layers requirements and is typically an ATM-based network. The architecture lacks well defined protocols for QoS negotiation as well as QoS mapping functions.

The specification of communication application requirements is generally done through the so called "flow specification". These flow specifications are a set of parameters, like the delay, the throughput and the jitter, that describe the communication needs for the transport of a given stream. [7] presents a recent proposal of a flow specification.

Considering existing works, there is a lack in application QoS management at end-systems and in the relation between applications needs and resources offers. We attempt to achieve this management on the base of a good expression, mapping and interpretation of applications QoS.

# 3  A FRAMEWORK FOR APPLICATIONS QoS MANAGEMENT

A link has to be made between the capabilities of the resources and the demands of multimedia applications. The first step towards achieving this link consists in providing a good QoS-based expression of the application needs. Then, it is important to define a model for interaction with the different resources involved in a multimedia communication session. The interaction model provides the resources with the application requirements based on application expressions. It allows also the support of dynamic QoS changes that may occur during applications execution. An analysis of applications requirements and their expression is first given. Then we describe the interaction model defined for the support of such requirements.

## 3.1 Multimedia requirements analysis and expression

A multimedia application handles a set of media, called also streams. These streams have several relations to each other and require a set of QoS guarantees from the system. The factors that have to be considered in studying applications requirements are the media types involved in the application and their transmission mode, the possible relations between these media and the dynamic QoS requirements. Suitable specifications should be defined to allow a simple interpretation by the system related to QoS mapping and management functions.

## Media types and transmission mode

A sending primitive is defined for media transmission, it allows the retreival of the QoS requirements of the medium to transmit. It has the form:

*Send(medium_type[parameters], sending_type[option])*.

The *parameters*, *sending_type* and *option* fields are optional. If they are not precised, a default value is assigned. the *parameters* field gives details on the medium type to transmit and allows the retrieval of the medium QoS requirements. An example of a medium type which describes a H.261 video type with p=1 corresponding to 64 Kbps and the QCIF format is:

*Video.H261[p=1,Frames/s=12,DimensionRatio=3/4,Lines=144,Pixels/line=192]*.

The *sending_type* may be *Asynchronous* (no time constraints), *RealTime* (a limit on maximum transmission delay) or *RealTimeInteractive* (a special case of a real-time transmission with stringent delay bound requirements, as for voice transmission in a conferencing application). The *option* field refers to the required degree (*loose*, *medium* or *tight*) defining the underlying constraint (table 1).

It is fundamental to note that, due to heterogeneous communication requirements for the streams of an application, the flows are transmitted on separate connections with differing QoS caracteristics.

## Media relations expression

Media of an application may have synchronization relations or logical relations. Synchronization relations describe temporal relations between media of an application which may have also different degrees depending on applications needs. The use of specifications like those proposed in [8] is a good way for describing the synchronization degrees. These synchronization degrees are: *optimal*, *good*, *acceptable* and *not tolerable*.

The *SynhronizedWith[option]* specification allows applications to express their synchronization requirements (*option* determines the required degree).

A logical relation may exist between streams of an application. For example, in a video conference, the image and the voice of a participant have a logical relation: voice can be presented without video, while the presentation of video without voice has no sense. Hence, in case of problems, the system should know which stream can be stopped without the need of stopping other streams. This is expressed by the specification *LogicalRelation[typeA; typeB->typeC->...]* (A and B may be presented without C, while the contrary is not possible). The notion of *group of streams* and a *number in the group* are introduced for this purpose.

## *Dynamic QoS*

Dynamic QoS changes can result from users interactivity or communication problems or lack of local resources. Degradations and improvements of QoS and media addition are considered as dynamic QoS issues. The possibility of degradation inside the Send primitive is shown in the following case: *Send{videoMPEG -> video.H261}* which means that if the system cannot support the transmission of the MPEG video type (requiring a throughput of about 1.5 Mbps), then it is possible to replace it by H.261 type which is less demanding in resources (H.261 with p=1 requires 64 Kbps).

## 3.2 A support model

An applications QoS manager (AQOSM) [9,10,11] is located at each end-system and is responsible of the management of applications QoS at this end-system. The AQOSM environment is composed of the applications, the local system resources including hardware (e.g. video codec and microphone) and software components (e.g. a synchronization service), and the interface to the communication sub-system. The interactions of the manager with its environment are achieved by *indications* that it receives and *notifications* it sends. Figure 1 shows the situation of the AQOSM. The dialogue between the different AQOSMs is achieved by the HLQNP protocol (High Level QoS Negotiation Protocol).

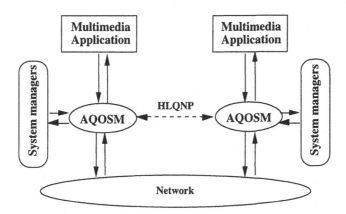

**Figure 1** The QoS management model

It is important to note that the AQOSM does not deal with resources management; however it has the task of providing the resources (system managers) with necessary information for their management. This information is retrieved from the application specifications.

## QoS mapping functions

The mapping functions required for the finding of QoS parameters that have to be requested by the AQOSM rely on Management Information Bases (MIBs) [12].

An example of a specific MIB entry is given in table 1. Different MIBs may exist for different systems. Other parameters contributing in QoS management are provided by applications like the *priority* parameters which help for management functions related to QoS degradations. In fact, each stream has two priorities: the *absolute priority* which is the priority of the application (to the other applications on the same site) to which it belongs and the *relative priority* which is the priority to other streams in the same application.

| Medium | Bandwidth | Delay | Jitter | Error rate | Hardware and software |
|--------|-----------|-------|--------|------------|----------------------|
| Voice.G722 | 64 Kbps | RealTime<br>[loose]: 1000ms<br>[medium]: 500ms<br>[tight]: 250ms<br>RealTimeInteractive<br>100ms | 10ms | $10^{-8}$ | G722 codec,<br>G722 driver |

**Table 1** A MIB entry example

## The interaction model

In order to achieve a local management at the end-system, we have developed an interaction model that describes the indications to and the notifications from the AQOSM.

The AQOSM can request a certain service from a resource (e.g. the allocation of an amount of memory from a memory manager) with a given QoS using the primitive *QoS.request(service, stream, QoS parameters)*. Changing the QoS of a service that has been obtained is done using *QoS.change(service, stream, QoS parameters)*. When a service is no more required, the AQOSM releases it using the primitive *QoS.release(service, stream, QoS parameters)*. The possibility to check for the availability of a certain QoS is offered to the manager by the primitive *QoS.check(service, stream, QoS parameters)* which gives information about the availability of the requested resources to satisfy the QoS parameters specified.

The resources reply to the manager requests by a set of primitives and may indicate, asynchronously, if a stream should be degraded or has to be released due to resource problems. After a QoS request from the manager, the concerned resource may accept the stream *QoS.accept(stream)*, accept it with degradation *QoS.modify(stream, degraded QoS parameters, reasons)* or refuse the request *QoS.refuse(stream, reasons)*. The defined interactions with the resources involve what we call *local QoS negotiation*.

One primitive is used by the manager to notify any problems or QoS changes to the application. This one is: *Notify(stream, problem, resource, parameters)*, which reports the nature of the problem (e.g. lack of resources), the concerned resource and stream and the related QoS parameters.

The above primitives do not offer the QoS guarantees, but provide the manager with a generic interface to ask for the application QoS requirements. For the QoS to be guaranteed, an effective ressource (local and network) reservation should be done. The local resources are mainly multimedia input/output devices, the memory and the CPU. The access to this latter has to provide real-time process scheduling. In our experimental implementation, we take the memory as an example of a resource that can be asked for the transmission of the streams.

# 4  THE DISTRIBUTED MANAGEMENT PROTOCOL

The distributed management functions are ensured by a protocol within the AQOSM, called HLQNP (High Level QoS Negotiation Protocol). HLQNP does not deal directly with resource management, but with application related management functions that may influence the resource management (e.g. asking for some amount of resources).

The role of HLQNP is the negotiation of applications QoS before applications execution and the dynamic QoS negotiations during the running of the applications. QoS negotiation allows the applications to agree on a common QoS, and its allowed degradation range. This negotiation is a management function that will provide the underlying network and system resources with the application requirements.

Based on the first applications QoS negotiation, the protocol allows the QoS follow-up and the renegotation of QoS parameters concerning media in transmission. These changes in QoS are the results of dynamic changes in applications or may be related to resource problems. We suppose that the underlying network is able to support dynamic QoS changes for the connections as HLQNP involves renegotiations.

In the following we present a quick overview of the protocol. The complete specifications of HLQNP are given in [11].

## 4.1 Protocol principles

HLQNP makes use of a reliable out-of-band facility for its message transmission. HLQNP does not apply error control on its messages. HLQNP ensures simplex negotiations, i.e. for one direction only. Duplex negotiations are considered as a special case.

Negotiations are done for single streams. If the application has several streams to transmit, separate negotiations have to be done. In fact, this is due to the fact that each stream will be transmitted on a separate connection with varying QoS characteristics.

Negotiations can be achieved under two modes: the *negotiated* mode allows a complete negotiation in the sense that the negotiation initiator should wait the reply from the receiver which can modify the QoS parameters; and the *forced* mode which is not really a negotiation and constitutes a simple QoS information report to the receiver which can only accept the QoS without any changes. In the latter mode, the sender may begin sending the stream without waiting for the receiver's reply. The negotiated mode is suitable for applications which are not affected by an initialization delay (due to the round-trip delay of the negotiations). The forced mode is suitable for dynamic and multicast negotiations or when it is sure that the receiver has sufficient resources to handle a stream. During a negotiation phase, only degradation of the desired QoS is allowed.

## 4.2 HLQNP description

### *The protocol messages*

Five messages are defined by HLQNP: NEGOTIATE, ACCEPT, MODIFY, REJECT and END.

The NEGOTIATE message carries QoS requirements and is initiated by the sending side of a stream. The ACCEPT message is initiated by the receiver of a NEGOTIATE message and reports the acceptance of the stream without any modification in QoS.

The MODIFY message allows the change by the receiver of a NEGOTIATE message of QoS parameters during a negotiation phase or the dynamic change of QoS (for a stream in transmission) if it is initiated outside of a negotiation.

The REJECT message reports the non acceptance of the stream with the specified QoS as a reply to a NEGOTIATE message. It is also used to report the abnormal termination of a stream (e.g. in case of resource problems during the stream transmission).

The END message is used for a normal termination of a stream. It allows the release of any resource reservation related to this stream. NEGOTIATE and MODIFY messages may be used in either modes of the protocol (negotiated or forced) during negotiation and renegotiation phases.

### *QoS negotiation scenarios*

A normal forced QoS negotiation is initiated after an application request (*Send* primitive). The AQOSM first applies the mapping function and determines the locally required resources which are pre-reserved (non confirmed reservation waiting for the result of the negotiation). It then invokes HLQNP which sends the corresponding NEGOTIATE message. The receiver of the message determines whether the local resources to receive the stream are available. In this case, it reserves these resources and replies with an ACCEPT message (or MODIFY message, if a degradation within the allowed range is necessary). Upon the receipt of the ACCEPT or MODIFY message, the sender confirms the resource reservation and the sending of the stream can start. If, in case of a lack of resources, the receiver cannot handle the stream, it will be refused before the negotiation and the application is notified (*Notify* primitive). If the receiver cannot accept the stream, a REJECT message is initiated and allows the sender to release the pre-reserved resources and to notify the application.

The NEGOTIATE message contains information about the application handling the stream, its priority, the stream direction, the stream mark indicating if it accepts degradation, information on the possible logical relation to other streams and the set of QoS parameters determined by the mapping function. Each parameter is given as a *parameter identifier*, a *desired value* and a *maximum degradation value*. In the negotiated mode, the receiver may change the desired value down to the maximum degradation value.

In forced mode, the sender can directly begin the stream transmission. We suppose in this latter case that the negotiation message will arrive to the destination before the data belonging to the stream. Only a REJECT message may be sent if the receiver cannot accept the stream's QoS; a degradation is not allowed.

Dynamic negotiations are achieved by the MODIFY message for existing streams. This message may be initiated by either the sender or the receiver. A MODIFY message is considered as a NEGOTIATE message. A renegotiation can be done in negotiated or forced mode. Introducing a new stream for an existing application involves a negotiation using the NEGOTIATE message.

## Local protocol information and algorithms

The protocol maintains information for each stream. This information is shown in table 2. The information on the streams is separated in two parts: one for the streams in transmission and the other for the streams in negotiation for which the resources are only pre-reserved.

| | |
|---|---|
| ApplicationIdentifier | StreamType |
| StreamIdentifier | StreamMark |
| Sending/Receiving Address | GroupMark |
| AbsolutePriority | GroupIdentifier |
| RelativePriority | InGroupNumber |
| NegotiationDirection | OriginalQoS |
| Allocated | Resources |

**Table 2** Local information on a stream

For each allocated resource the table contains the amount reserved for the stream (e.g. amount of memory). This information allows the protocol to decide if it is possible, by degrading or removing the stream, to serve another stream with higher priority.

In all negotiations (forced or negotiated, dynamic or not), it is possible to stop a stream with lower priority in order to accept a stream with a higher priority. This issue is considered by different algorithms defined by the protocol [11].

If a stream requires an amount of resources which is not available, other streams with lower relative priority and accepting degradation within the same application may be degraded or removed (depending on the amount of resources they have) to serve the new stream. If this does not offer a proper solution, streams belonging to other applications with lower absolute priority may then be degraded or removed. When a stream is removed, the protocol checks if it has a logical relation to other streams and send a REJECT message for these streams (depending on the InGroupNumber field which decides if the stream has to be removed or not).

# 5 CONCLUSION

We have presented a framework for the management of applications QoS at the end-systems. Based on the definition of applications requirements and an interaction model with the resources, the framework allows an efficient distributed management of QoS on behalf of the applications. A description of the HLQNP management protocol has been given. HLQNP allows the QoS negotiation between the applications to provide the underlying resources with their requirements.

We are currently integrating the framework within an architecture comprising four levels: application, synchronization, transport and network [13]. Extensions to HLQNP are required for this purpose. They are related to the definition and the negotiation of *QoS profiles* providing information for the choice of suitable communication services and not only QoS parameters.

# REFERENCES

[1]    ITU-TS recommandation I.352, "Network performance Objectives for Connection Processing Delays in an ISDN", 1988.

[2]    Danthine, A., Baguette, Y., Ledu, G., and Léonard, L., "The OSI95 Connection-Mode Transport Service-The Enhanced QoS", Fourth IFIP Conference on High Performance Networking, Liège, Belgium, December 14-18, 1992.

[3] "Experimental Internet Stream Protocol, Version 2 (ST-II)", Network Working Group, RFC 1190, IEN-119, CIP Working Group, C. Topolcic, Editor, October 1990.

[4] Zhang, L., Deering, S., Estrin, D., Shenker, S. and Zappala, D., "RSVP: A New Resource ReSerVation Protocol", to appear in IEEE Network Magazine.

[5] Banerjea, A., and Mah A., "The design of a Real-Time Channel Administration Protocol", Second International Workshop on Network and Operating System Support for Digital Audio and Video, Heidelberg, Germany, November 18-19, 1991.

[6] Leopold, H., Campbell, A., Hutchison, D., and Singer, N., "Towards an Integrated Quality of Service Architecture (QOS-A) for Distributed Multimedia Communications", Fourth IFIP Conference on High Performance Networking, Liège, Belgium, December 14-18, 1992.

[7] Partridge, C., "A Proposed Flow Specification", Network Working Group, RFC 1363, September 1992.

[8] Steinmetz, R., "Synchronization Properties in Multimedia Systems", IEEE Journal on Selected Areas in Communications, Vol. 6, No. 3, April 1990.

[9] Tawbi, W., Fédaoui, L., and Horlait, E., "Management of Multimedia Applications QOS on ATM Networks", IFIP International Conference on Computer Networks, Architectures and Applications (Networks'92), Trivandrum, India, October 26-28, 1992.

[10] Tawbi, W., Fédaoui, L., and Horlait, E., "Dynamic QoS Issues in Distributed Multimedia Systems", Second International Conference on Broadband Islands, Sponsored by IEEE, Eurobridge and CEE, Athens, Greece, June 14-16, 1993.

[11] Tawbi, W., "Quality of Service in Multimedia Communication Systems: A study of a Framework and Specification of a Negotiation Protocol between Applications", Ph.D Thesis, Université Pierre et Marie Curie, Paris VI, Paris, France, December 1993 (in english).

[12] Rose M.,"The Simple Book - An Introduction to Management of TCP/IP-based Internets", Prentice Hall, 1991.

[13] Besse, L., Dairaine, L., Fédaoui, L., Tawbi, W. and Thai, K., "Towards an Architecture for Distributed Multimedia Applications Support", IEEE ICMCS Conference, Boston, USA, May 15-19, 1994.

# 13

# SUPPORTING CONTINUOUS MEDIA APPLICATIONS IN A MICRO-KERNEL ENVIRONMENT

G. Coulson, G.S. Blair, P. Robin and D. Shepherd

*Department of Computing, Lancaster University,*
*Lancaster LA1 4YR, U.K.*
*e.mail: mpg@comp.lancs.ac.uk*

## ABSTRACT

Today's operating systems were not designed to support the end-to-end real-time requirements of distributed continuous media. Furthermore, the integration of continuous media communications software into such systems poses significant challenges. This paper describes a design for distributed multimedia support in the Chorus micro-kernel operating system environment which provides the necessary soft real-time support while simultaneously running conventional applications. Our approach is to extend existing Chorus abstractions to include QoS configurability, connection oriented communications and real-time threads. The paper defines a low level API for distributed real-time programming and describes an implementation which features a close integration of communications and thread scheduling and the use of a split level scheduling architecture with kernel and user level threads.

## 1 INTRODUCTION

A considerable amount of research has already been carried out in communications support for continuous media over high speed networks. However, much less work has been done in the area of general purpose operating system support for continuous media. Typically, end system implementations have either been embedded in non real-time operating systems such as UNIX and suffered from poor performance, or have been implemented in specialised hardware/software environments unable to support general purpose applications.

The SUMO Project at Lancaster [1] is addressing this deficiency in the state of the art by extending a commercial micro-kernel (i.e. Chorus [2]) to support continuous media applications alongside standard UNIX applications. Chorus is a useful starting point for continuous media support as it includes a number of desirable real-time features. However, in common with other micro-kernels, it fails to adequately support continuous media in a number of key areas. First, communication in Chorus is *message based* whereas continuous media requires *stream-oriented* communications.

215

Second, Chorus offers no *quality of service* (QoS) control over communications and only coarse grained relative priority based scheduling to control the QoS of processing activities. Finally, Chorus does not offer *end-to-end* real-time support spanning both the communications and scheduling components.

To overcome these deficiencies we introduce the concept of a 'flow'. Flows characterise the production, transmission and eventual consumption of a single media stream as an integrated activity governed by a single statement of QoS. Realisation of the flow concept demands tight integration between communications, thread scheduling and device management components of the operating system. It also requires careful optimisation of control and data transfer paths within the system.

The rest of this paper is structured as follows. Section 2 provides the background on the Chorus micro-kernel necessary to understand the rest of the paper. Then section 3 describes the programming interface to our multimedia facilities. Following this, section 4 discusses the implementation of the API, concentrating on communications and scheduling issues. Finally, section 5 presents our conclusions and indicates our plans for future work.

## 2 BACKGROUND ON CHORUS

Chorus, conceived at INRIA, France, is a micro-kernel based operating system which supports the implementation of conventional operating system environments through the provision of 'sub-systems' or 'personalities' (for example a sub-system is available for UNIX SVR4). The micro-kernel is implemented using modern techniques such as multithreaded address spaces and inter-process communication with copy-on-write semantics. The basic Chorus abstractions are *actors*, *threads* and *ports*, all of which are named by globally unique and globally accessible unique identifiers. Actors are address spaces and containers of resources and may exist in either user or system space. Threads are units of execution which run code in the context of an actor. By default, they are scheduled according to either a pre-emptive priority based scheme or round robin timeslicing. Ports are message queues used to hold incoming and outgoing messages. They can be aggregated into *port groups* to support multicast messaging and may be migrated between actors. Inter-process communication is datagram based and supports both request/reply messages (via the ipcCall() and ipcReply() system calls and one shot messages (via ipcSend() and ipcReceive()).

Chorus has several desirable real-time features and has been widely used for embedded real-time applications. Its real-time features include pre-emptive scheduling, page locking, system call timeouts, and efficient interrupt handling. Chorus also incorporates a framework, called *scheduling classes*, which allows system implementers to add new scheduling policy modules to the system. These modules are upcalled each time a scheduling event occurs. Modules impose their scheduling decisions by manipulating a global table of thread priorities.

Unfortunately, Chorus' real-time support is not sufficient for the requirements of distributed multimedia applications, principally because there is no support for QoS control and resource reservation:-

- although it is possible to specify thread scheduling constraints relative to other threads, absolute statements of requirement for individual threads cannot be made,
- the exclusive use of connectionless communications makes it impossible to pre-specify communications resource allocation.

In addition, Chorus suffers from a lack of communications/ scheduling integration. This means that there is no way to provide timely scheduling in concert with communications events as required by end-to-end continuous media communications. Note, however, that the above limitations are not unique to Chorus: they are shared by most of the other micro-kernels in current use (e.g. [3, 4]).

# 3 PROGRAMMING INTERFACE AND ABSTRACTIONS

To remedy its current deficiencies for real-time continuous media support and real-time control, we have extended the Chorus API with new low level calls and abstractions (provided in a user level library called *libflow*).

## 3.1 Primitive Abstractions

The primitive abstractions, described in the forthcoming sections, are as follows:-

- *rtports* - communication end points for real-time communications,
- *devices* - hardware and/or software producers, consumers and filters of real-time data,
- *rthandlers* - user defined procedures which manipulate real-time data coming from or going to an rtport,
- *QoS controlled connections* - communication channels with a specific QoS,
- *QoS handlers* - user defined procedures which are upcalled when QoS commitments degrade, and
- *threads* - a set of real-time lightweight thread management and synchronisation primitives.

## *Rtports and Devices*

Rtports are an extension of standard Chorus ports and serve as end-points for both continuous media communications and real-time messages with bounded latency. As such, they participate in the implementation of end-to-end flows (in conjunction with rthandlers and QoS controlled connections). Like Chorus ports, rtports are named globally and can be accessed in a location independent fashion from anywhere in the distributed system.

There are, however, the following differences between Chorus ports and rtports:-

- rtports have an associated QoS,
- the internal buffers of rtports can be directly accessed by the application programmer without the overhead of a system call,
- rtports may not migrate because the QoS commitments offered by the rtport assume a fixed association between the underlying device and the actor performing the create operation.

A *device* is a producer, consumer or filter of real-time data which supports the creation of rtports. Devices may be either software drivers for physical devices or independent software objects that generate, process or consume continuous media data. One important type of device is the *null* device. This 'device' enables pieces of user code to act as data sources and sinks. An rtport associated with such a device is similar to a conventional Chorus port except that ipcSend() and ipcReceive() calls made on such a port may be latency bounded if the rtport is the end-point of a QoS controlled connection.

Rtports are created with the following call:-

```
typedef enum {s_message, s_stream} flow_type;
status rtportCreate(DEV *d; flow_type service; rtport *p);
```

The user specifies to the system a device on the local site with which the new rtport should be associated, the required QoS of the rtport and the type of service required: either a QoS controlled message service or a stream service. The device argument is a standard Chorus unique identifier which refers to a hardware or software device. Note that rtportCreate() will only succeed when the referenced device resides on the local site.

The QoS specification is used to denote the potential level of service of the rtport and to deduce future resource allocation needs. However, resources are not actually committed until the rtport is involved in a connection (see later sub-section). QoS parameters are specified to the system by means of a data structure called a QoSVector. Different definitions of this data structure are used depending on the service type required. QoSVector definitions for the QoS controlled message and the stream service are given below :-

```
typedef enum {best_effort, guaranteed} com;
typedef enum {isochronous, workahead} del;
                                        typedef struct {
typedef struct {                            com commitment;
    com commitment;                         int buffsize;
    int buffsize;                           int priority
```

```
int priority;                        int latency;
int latency;                         int error;
int error;                       } MessageQoS;
int buffrate;
int jitter;                      typedef union {
del delivery;                        MessageQoS mq;
int error_interval;                  StreamQoS sq;
} StreamQoS;                     } QoSVector;
```

The first four parameters are common to both service types. The *commitment* parameter allows the programmer to express a required degree of certainty that the QoS levels requested will actually be honoured. If the commitment is *guaranteed*, resources are permanently dedicated to support the requested QoS levels. Otherwise, if the commitment is *best effort*, resources are not permanently dedicated and may be preempted for use by other activities. A later sub-section describes special reporting facilities provided in the case that requested QoS levels are violated. *Buffsize* specifies the required size of an internal buffer to be associated with the rtport. *Priority* is used to control resource pre-emption for connections. All things being equal, a connection with a low priority will have its resources pre-empted before one with a higher priority. *Latency* refers to the maximum tolerable end-to-end delay, where the interpretation of 'end-to-end' is dependent on whether rthandlers are attached to the rtport (see later in this section).

The *error* parameter has a different interpretation depending on the type of service requested. For stream connections, error, which is used in conjunction with *error_interval*, refers to the maximum permissible number of buffer losses and corruptions over the given interval. In the case of message connections, error simply represents the probability of buffers being corrupted or lost (error_interval is not applicable to message connections).

For the stream service only, there are three additional parameters. *Buffrate* refers to the required delivery rate of buffers at the sink end of the connection. *Jitter*, measured in milliseconds, refers to the permissible tolerance in buffer delivery time from the periodic delivery time implied by buffrate. *Delivery* also refines the meaning of buffrate. If delivery is *isochronous*, the stream service delivers precisely at the rate specified by buffrate; otherwise, it attempts to 'work ahead' (ignoring the jitter parameter) at rates temporarily faster than buffrate. One use of the workahead delivery mode is to support applications such as real-time file transfer. Its primary use, however, is for pipelines of processing stages.

## Rthandlers

Rthandlers are user supplied C functions which may (optionally) be attached to rtports. They may be attached to both sending and receiving rtports. Rthandlers are upcalled from their associated rtport whenever data is required at a source rtport or has been delivered, by a connection, to a sink rtport. The thread which upcalls the rthandler runs in user mode and thus allows the user the freedom to provide arbitrary code for rthandlers. The upcalling of an rthandler performs two logically distinct functions:-

   i) *event notification* - it is indicated that data is required or has been delivered, and

   ii) *data transfer* - it is made possible for the rthandler to access the rtport's buffer to insert or extract data as appropriate.

   Applications can use rthandlers either for the notification of events alone, or for both event notification and data transfer. We feel that this separation of notification and delivery is important for continuous media applications. It permits applications to choose whether they want to actively process continuous media data in user space, or merely to track the passage of continuous media generated and consumed in supervisor space. This latter case arises when the device under consideration is, for example, a kernel managed video device with associated frame buffer which is receiving data directly from the network card. Here, efficiency can be maximised as continuous media data need not cross protection domains.

   The call to attach an rthandler to a rtport is as follows:-

```
typedef int(Rthandler)(Buffer **b; int *size; int *event;
                       time_t *send_timestamp, *recv_timestamp;
                       bool admission);
status rtportAttachRthandler(rtport *p; Rthandler f,
                             int eventmask; short priority);
```

   The first two arguments to rthandler functions inform the application's code of the size and address of the rtport's internal buffer. In cases where the buffer is mapped into user space, this permits user code to directly supply/extract data from the buffer while rthandlers are executing. Rthandlers can assume that they have exclusive access to the buffer for as long as their call is extant. In the case of kernel supported devices, direct user access to buffers may or may not be possible depending on the protection attributes imposed by the device and/or the kernel. If access to buffers is disallowed (as indicated by a NULL value for the b parameter), the rthandler performs an event notification but not a data transfer role.

   The third parameter, *event*, allows application code to associate multiple logical message types with a single rthandler. Source rthandlers provide an integer event identifier and this is passed on, unchanged and uninterpreted, with the corresponding buffer, to the destination rthandler. The fourth and fifth arguments supply timestamps for the benefit of the receiver. These relate to the times at which the buffer was obtained at the source and delivered at the sink. Finally, the *admission* parameter is used to allow the infrastructure to perform a scheduling admission test by determining the execution time of an rthandler. When it sees a *true* admission value, application code in an rthandler is expected to direct the calling thread on a dummy run through a 'typical' execution path so that the resource manager can derive an estimate of the execution time of the rthandler in normal circumstances. This execution time is added to the a priori known time for protocol processing to help derive the deadline of the rtport's associated thread. Note that admission is only given as *true* under these circumstances; at all other times it is given as *false*.

The first two parameters to the rtportAttachRthandler() call itself specify the rtport to which the rthandler is to be attached and the rthandler function. The third and fourth parameters are used to control the behaviour of rthandlers when multiple rthandlers are attached to the same rtport. *Eventmask* is a bitmap used to determine for which set of events (see above) this particular rthandler will be called, and *priority* is used to determine the order in which multiple rthandlers are called when more than one share a common event. Finally, the rtportAttachRthandler() call will fail if the given rtport was not created in the current address space. This is because the virtual address of the rthandler must have an interpretation in the rtport's supporting address space.

Rthandlers are central to our design for two major reasons:-

* *programmability*
  Because rthandlers are executed under QoS constraints imposed by their associated rtport, they provide a way to execute application code without compromising the system provided QoS support required for the realisation of flows. We also contend that structuring the API with rthandlers is a natural and effective model for real-time programming. Real-time programming is considerably simplified when programmers can structure applications to react to events and *delegate to the system* the responsibility for initiating events. Of course the programmer is still ultimately in control of event initiation but this control is expressed declaratively through the provision of a QoSVector parameter and need not be explicitly programmed in a procedural style.
* *efficiency*
  An efficiency gain results from the use of a single thread (originating in the communication system) for both protocol and application processing. In conventional systems, applications interface with communications by performing system calls which block and reschedule if the protocol is not ready to send or if data has not yet arrived. With the rthandler implementation, on the other hand, no context switch is incurred and it is not necessary for the application and protocol to wait for each other as the protocol always initiates the exchange and the application code should always be ready to run.

Figure 1 illustrates devices, rtports and rthandlers. It also shows incoming QoS controlled connections (shown as heavy black lines), one for each of the two rtports, and rthandlers attached to the rtports and upcalling into user space. One of the rtports is associated with a *null* device (where the rtport is implemented in the libflow user level library), and the other with a physical device (implying an rtport implementation in supervisor address space).

In the case of the actor rtport, the rthandler is performing the role of both event notification and data transfer. In the case of the kernel device rtport, the rthandler plays a similar event notification role (i.e. data is being sent/ received), but application code does *not* directly participate in the data transfer. Instead, data is directly obtained from/ delivered to the device by the connection associated with the device, and need not cross into user space.

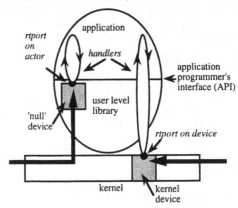

**Figure 1** Devices, Rtports and Rthandlers

## *QoS Controlled Connections*

All communication in standard Chorus is connectionless and datagram based. However, as noted in section 2, flow services require resource reservation commitments both in the end-system and in the network. Because of this, we have added simplex connection oriented communications called QoS controlled connections to abstract over the necessary resource allocation. The call to set up such a connection is as follows:-

```
status rtportConnect(rtport *source, *sink; QoSVector *qos;
                     rtport *ctl);
```

The rtportConnect() call takes two rtports, a source and a sink, as its primary parameters. Notice that because of the location independent nature of rtports, it is possible to call rtportConnect() from a site entirely separate from the sites on which the source and sink rtports reside. This is a convenient facility for distributed multimedia applications which are often structured as a centralised master process supervising and controlling a number of physically distributed sources and sinks [5]. The internal protocol for this facility, which we call *remote connection*, is fully described in [6].

The QoS parameters are the same as those used at rtport creation time except that the values specified at connect time are *actual* rather than *potential.* The values specified must be less than or equal to those specified in the QoS of both rtports involved, or rtportConnect() will fail. The two rtports must also have been created with the same flow service type: either message or stream types. The *latency* QoS parameter subsumes rthandler execution time if rthandlers are attached and thus specifies the full latency of an application-to-application flow. If rthandlers are not attached, latency is interpreted as rtport to rtport. In the latter case, latency is measured from the time a thread calls ipcSend() to the time the buffer is received (but not necessarily delivered to the application) at the sink rtport.

There are two categories of connection corresponding to the two types of flow supported by rtports:-

- *message connections* - these wait passively until activated by ipcSend() calls in the conventional manner. It is possible to attach rthandlers at the sink end of message connections, but source rthandlers are inapplicable because there is no active entity in the connection to call them.
- *stream connections* - each end of a stream connection tries to *actively* obtain/ deliver data at the rate determined by buffrate. If rthandlers are attached, this results in the calling of the rthandlers, otherwise the connection blocks until the application calls ipcSend() or ipcReceive().

The final argument to rtportConnect() is a result parameter which returns a new rtport used to dynamically control the behaviour of connections. The 'operations' available on the control rtport are the following:-

- *renegotiate* - this allows the user to dynamically change the QoS of the connection by supplying a new QoSVector argument,
- *disconnect* - to destroy a connection,
- *start, stop* - respectively activate and de-activate the connection,
- *prime* - ready the end-to-end connection by filling the receive buffers so that a subsequent start will take immediate effect.

The last three operations are only applicable to stream connections. Note that start, stop and prime can also be used for cross-stream synchronisation purposes; their use in this context is described in [6].

In implementation, connection establishment is provided by a per-site connection manager which is realised as a user level actor. The connection manager maintains a list of actors and their associated rtports for that site. The manager accepts incoming connection requests and dispatches them to a listening lightweight thread in the appropriate rtport implementation.

## *QoS Handlers*

QoS handlers are upcalled by the system in a similar way to the rthandlers described above. However, whereas rthandlers notify communication events and allow access to communication buffers, QoS handlers are used to notify the application layer when the QoS commitments provided by connections have been violated. The intention is that QoS handlers will usually be attached and supported by application level QoS manager objects. QoS managers embody a particular policy for coping with QoS degradations. For example, they may attempt to request renegotiation of QoS or choose a connection to close down on the basis of application defined criteria. The call to attach a QoS handler is:-

```
typedef int (QOSHandler) (QoSVector *current, *new);
status rtportAttachQOSHandler(rtport *p; QOSHandler f);
```

## *Threads*

Although lightweight threads (see section 4.1) are implicitly created on behalf of applications when they establish QoS controlled connections, we also allow applications to *explicitly* create lightweight threads. Our thread types and thread creation primitives are similar to those described in [7] and hence will not be described here. However, our thread synchronisation primitives are novel in that they incorporate the concept of *deadline inheritance*. This allows a thread requiring a subcomputation to be performed by a second thread to pass on its deadline to the second thread so that deadlines can be attached to entire logical computations as well as just individual threads.

The central thread synchronisation calls, which are based on *eventcounters* and *sequencers* [8], are as follows :-

```
void advance(eventcounter *e, bool inherit);
bool await(eventcounter *e; u_long target; time_t timeout);
```

Note that the boolean result of await() is used to distinguish whether the call returned due to timeout expiry or the eventcounter target being reached. The semantics of deadline inheritance are as follows:-

- Each thread maintains a list of deadlines sorted earliest deadline first. The entry at the front of the list is the *effective deadline* used by the scheduler. When a thread starts to run it has only one entry in its list.
- Whenever a thread T executes advance(e) with the inherit flag set, any threads that were blocked on an earlier call of await(e), and freed as a result of the advance(e) call, *inherit* the effective deadline of T *iff* T has an earlier deadline than their current effective deadline. The mechanism of inheritance is to place the inherited deadline at the front of the deadline list of the inheriting thread. When multiple threads share the same deadline value, the one with the shortest deadline list takes precedence.
- Whenever a thread which has passed on a deadline to other threads decides to terminate, all inherited entries associated with the terminating thread are removed from the deadline lists of all the

inheriting threads. This may cause a change in the effective deadline of one or more of the inheriting threads.

- The programmer can explicitly enable or disable inheritance by appropriately setting the second parameter of advance().

Note that deadline inheritance is *not* intended to address the related well known problem of *priority inversion* (e.g. see [7]) whereby a thread with a late deadline holds a mutual exclusion lock for which a thread with an early deadline is waiting. Rather, it is used to release application programmers from the need to structure their event handling code (expressed in rthandlers) in a single sequential thread. With deadline inheritance, code can be structured in terms of an arbitrarily complex system of concurrent worker threads without compromising the predictable performance of rthandlers.

## 3.2 Compound Abstractions

The compound abstractions, described in this section, are as follows:-

- *invocation* - a message based request/ reply service
- *pipeline* - concatenations of QoS controlled stream flows containing intermediary processing stages.

## *Invocation*

The invocation service is a compound service realised in terms of two message connections arranged in a back to back, request/ reply, configuration.

To provide a convenient interface to the programmer, the libflow library exports abstractions which relieve the programmer of the necessity of explicitly manipulating the two message connections. These abstractions comprise an operation to create an invocation service instance (invocationConnect()) and operations to send and receive data on an invocation service instance (ipcCall() and ipcReply()). The operation to establish an invocation service instance is as follows:-

```
status invocationConnect(rtport *clientPrt, *serverPrt;
                         InvocationQoS *qos; rtport *ctl);
```

InvocationConnect() uses the supplied rtports (clientPrt and serverPrt) as the endpoints of the request connection. The underlying implementation transparently allocates the two rtports required for the reply connection. These two rtports do not need to be visible to the user because of the semantics of the ipcCall() and ipcReply() calls (see below). Note, however, that it is the programmer's responsibility to attach an rthandler to serverPrt before using the service. The InvocationQoS parameter denotes the required end-to-end QoS of the invocation and includes fields such as round-trip latency and an option field which specifies at-most-once or at-least-once semantics. This information is used to determine the QoS parameters in the two underlying connections.

In use, a client thread at the initiating end calls ipcCall(). Then, at the server end, the incoming call is executed on the sink rtport's rthandler. When replying, the server calls ipcReply() without explicitly specifying an rtport.

## Pipelines

Our design supports the requirement for pipelines through the concatenation of stream connections. Processing stages in pipelines are realised as rthandlers attached to intermediate *filter* devices. Some filter devices, e.g. compression functions, are implemented in hardware and managed by the kernel. Other filtering requirements, however, can be met by application level software processing. To permit application writers to implement such processing, we provide a generic software filter device known as a *connector*. Connectors are implemented in the libflow user level library in a similar way to the invocation abstractions described above. Figure 2 illustrates a three stage pipeline with two intermediate connector devices.

The following call is used to create a connector:-

```
status rtconnectorCreate(DEV *d);
```

A connector device is an encapsulated bounded buffer on to which programmers can create rtports. Programmers create a single rtport on the connector to serve as both the sink of the upstream connection and the source of the downstream connection. An rthandler is then attached to the rtport to encapsulate the required application processing. The rthandler is invoked from below when a buffer is available on the upstream connection and the downstream connection waits until the rthandler returns. As the connector's buffer is shared between the two connections, no buffer copy overhead is incurred between pipeline stages.

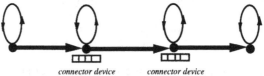

connector device          connector device
**Figure 2** Pipeline

The following call enables pipelines to be created with a single statement of end-to-end QoS (and thus fall within the definition of flow):

```
status pipelineConnect(rtportList rtports; StreamQoS *qos;
                       rtport *ctl);
```

PipelineConnect() takes a list of rtports, all of which are assumed to have an rthandler already attached. An ordinary StreamQoS structure specifies the end-to-end QoS. If the *delivery* flag in the StreamQoS structure is set to isochronous, only the last connection in the pipeline is actually set up as an isochronous binding. This is because it is desirable to permit as much asynchronicity as possible so that the pipeline can run with minimal constraints and maximal elasticity. As long as the last pipeline stage runs with the required isochronicity the users end-to-end QoS specification is sufficiently upheld.

In implementation, the end-to-end QoS specification supplied to pipelineConnect() is partitioned across the various component connections using a resource reservation protocol described in [9]. The result parameter *ctl* is exported by libflow as a convenient handle to control the end-to-end pipeline flow. When programmers invoke control operations on this rtport, libflow must perform a (non trivial) mapping to the control rtports of the full set of constituent connections.

# 4 IMPLEMENTATION

## 4.1 Scheduling

The scheduling implementation exploits the concept of *lightweight threads* to minimise the overhead due to context switches. Lightweight threads are implemented in the libflow library and multiplexed, non timesliced, on top of Chorus kernel level threads acting as 'virtual processors' (VPs).

When used to implement stream connections, lightweight threads run *periodically* at a rate determined by the QoS of the connection. In our current scheme, threads belonging to connections with guaranteed commitment and isochronous delivery are scheduled according to an extended preemptive earliest deadline first (EDF) [11] scheme which honours jitter bounds [10] (an admission test is used at connect time to ensure that sufficient resources are available). Threads with other combinations of commitment and delivery use traditional EDF (also with an appropriate admission test).

The implementation architecture for real-time thread scheduling is a split level scheme [12] consisting of a single kernel scheduler (KLS) and multiple co-operating user level thread schedulers (ULS), one in each actor. Each actor multiplexes lightweight threads on a small number of VPs dedicated to the actor (ideally only one VP for uni-processors).

For EDF threads, the scheme maintains the following invariants with respect to the two types of scheduler:-

i)   each ULS runs the lightweight thread in its actor with the earliest deadline,

ii)  the KLS runs the virtual processor of the actor with the globally earliest lightweight thread deadline.

The necessary information exchange between the kernel scheduler and the user level schedulers is accomplished via a combination of shared memory and upcalls from the kernel [12]. In the implementation of the shared memory area, each VP has an associated context structure [13] which contains information such as current lightweight thread context, next runnable thread and earliest deadline of a lightweight thread supported by this virtual processor. Each context is mapped into kernel space and used to exchange/share information with the kernel, thus avoiding unnecessary system calls. For example, the system clock is mapped to each virtual processor (read only) avoiding the cost of a system call to get the time value. The globally earliest deadline of each virtual processor is read on each kernel level rescheduling operation by the KLS to compute the next virtual processor to schedule.

## 4.2 Communications

### Fundamental Mechanisms

It is essential that VPs are able to communicate with optimal efficiency with parts of the system outside the context of their own actor. For example, data arrivals from the network must be notified to VPs in as efficient a manner as possible and VPs must be able to efficiently communicate with other VPs and kernel services. To meet these needs we have designed special purpose upcall and downcall mechanisms. For upcalls, we reject the standard Chorus strategy of having a thread waiting on a port because of the associated overhead of a synchronisation and context switch. Instead, we employ a *software interrupt* mechanism whereby the occurrence of an event causes the virtual processor to jump to an entry point in its user level scheduler. When this happens, the user level scheduler saves the current lightweight thread context and schedules the appropriate lightweight thread to perform the required action.

For downcalls, we use a mechanism known as an *asynchronous systems calls*. To perform an asynchronous system call, the user level places some parameters in the kernel/ user shared memory area and then sets a special byte, also in the shared area, to the required call's op code. The kernel inspects these bytes on each system clock tick and, if it finds any user actor's op code byte set, it wakes a system thread to perform the required service. The result of the call is eventually be notified by a software interrupt as discussed above. Unlike a traditional system call, this mechanism incurs minimal context switch and domain crossing overhead.

### Remote Communications

Connections between devices on different machines are implemented via a connection oriented communications stack specifically designed to support

QoS controlled communications [14]. As recommended by researchers in the area (e.g. [16]), our design uses no multiplexing in the protocol stack above the link layer. The transport protocol supports QoS parameterisation at connection set-up time and permits dynamic renegotiation of QoS levels on open connections. The protocol uses a rate based scheme [15] for flow control whereby sources and sinks negotiate a mutually acceptable transfer rate at connection set-up time. This allows data to flow at a smooth rate, which is important for continuous media, and also permits responsive back pressure to be applied when the sink runs out of buffers. Rate based flow control has further advantages. First, is lightweight and permits higher throughput than schemes based on windowing. Second, it decouples flow control from error control so that connections which do not require error control do not need to pay for it. Third, it fits in nicely with the notion of periodically schedulable threads.

The transport protocol can run in either supervisor mode or user mode. Supervisor mode is appropriate for connections involving kernel supported physical devices as both data transfers and execution for these connections are confined to supervisor memory space. User mode is appropriate for connections terminated by user supplied rthandlers attached to null devices. In the user mode implementation, the transport protocol runs in the context of lightweight threads in the address space of a user virtual processor, the network card is accessed through downcalls to a device driver which provides an interface at the link-layer, and the protocol itself is implemented in the libflow user level library. It is essential for efficiency that the user mode protocol implementation minimises the number of downcalls per user level buffer. To achieve this, the implementation batches data to reduce the number of transfers to the network card and also minimises memory allocation calls by maintaining its own buffer cache. Experience will tell if these techniques are sufficient but results reported in [17, 18] are encouraging.

Note that the user library transport implementation and rthandler mechanism simplify the task of scheduling in two respects. First, the deadline and required CPU time for the processing of each continuous media buffer is known in advance. The former is obtained implicitly from the QoS specification of the connection; the latter is deduced by combining the transport and rthandler execution times. Second, the rthandler scheme eliminates any need for synchronisation between the application and a distinct transport entity: a single seamless thread of execution subsumes both these activities.

To realise the active semantics of stream connections, connections have dedicated lightweight threads at each end (in the case of connections whose end rtports are kernel managed, these lightweight threads are supported by virtual processors running in supervisor mode in supervisor actors). The source thread is responsible for continually executing the user's rthandler(s), obtaining data (either from the rthandlers or directly from a physical device as appropriate), and executing the transport protocol. The sink thread operates analogously; it is awakened when the full set of link-layer packets making up a user level buffer have been received, and then executes the transport protocol and delivers the data either by calling the user's rthandler(s) or by delivering data directly to a device.

## Local Optimisations

A number of important optimisations are possible if communication is between devices in the same address space. In this case, connections between rtports in the same actor are simply implemented as a single lightweight thread which repeatedly calls the source rthandler(s) with a particular buffer address and then calls the sink rthandler(s) with the same address. This single mechanism serves to implement both the data transfer and the delivery notification aspects of the communication. It also implicitly ensures mutual exclusion to the buffer by the source and sink rthandlers. Connections between rtports in the same actor are often used in the context of intra-actor pipelines. To minimise data copying in such pipelines, the address of a single buffer is passed from stage to stage as the various stages of the pipeline are executed. Finally, when the last pipeline stage in the actor has disposed of the data, the buffer is released. If new data arrives at the actor while the previous data is being passed along the pipeline, processing of this data proceeds concurrently using a separate buffer. In this way, it is possible to efficiently implement arbitrarily long intra-actor pipelines without incurring copying overheads.

Further optimisations are possible in communications between kernel and user space. Chorus currently incurs (at best) one copy and one virtual memory remap for a data transfer from the network driver to user space. We can reduce this to zero overhead per transfer and one single virtual memory remap incurred at connection establishment time. Our strategy is to dedicate shared, per connection, physical memory buffers between the network driver and user space which are mapped in the address spaces of both. Note that, by a simple extension, this scheme can also be applied to the case of actor to actor communications on the same machine. The strategy here is for each actor to unmap a buffer from its own address space, map it into the address space of the next actor in the pipeline, and then pass a software interrupt to the next actor to implement the event notification aspect of the connection. This whole sequence of operations is encapsulated in a single specialised software interrupt upcall. These topics, together with other issues concerning the allocation and preemption of physical buffers to/ from user programs are discussed in [10].

References [10, 1] contain more complete details of the implementation as a whole.

# 5 CONCLUSIONS

We have presented a low level API and implementation scheme for distributed multimedia support in a Chorus based micro-kernel environment. As the basic abstractions of Chorus are comparable to those of other current micro-kernels such as Mach and Amoeba we expect that it should be possible to extend other micro-kernels in a similar way.

At the present time, we have established an experimental infrastructure consisting of two 80386 based PCs running Chorus. The PCs are equipped with VideoLogic audio/ video/ JPEG compression boards. The machines are now equipped with ATM interface cards and connect to a local ATM network. This is enabling us to extend our investigation of QoS issues and resource reservation to the network and work on an overall architecture for QoS. Our plans for establishing an ATM based infrastructure and end-to-end QoS architecture are detailed in [19].

We are currently working on the implementation of both the transport and scheduling aspects of the design presented in this paper. The scheduling implementation is based around the Chorus scheduling classes facility mentioned in section 2. This provides an ideal implementation basis for our kernel level scheduler. The user level scheduling and thread support aspects of the design are based on a simple non-timesliced package which we are extending with a shared memory interface to the new kernel scheduling class and upcalls and downcalls. As mentioned above, our transport implementation is based on a pre-existing protocol. We are currently porting this to Chorus from the transputer based platform for which it was initially designed.

There remain a number of important issues which we have not yet addressed. One involves the extension of connections and rtports to operate in the context of port groups as supported by standard Chorus. This is non trivial due to the inherent problems of connection oriented multicast [20]. A second issue is the provision of a higher level distributed programming platform which we are currently investigating in co-operation with researchers at CNET, France. The platform will be based on the ISO's emerging standards for Open Distributed Processing with extensions for real-time synchronisation, continuous media and QoS [21].

## Acknowledgement

The research reported in this paper was funded under UK Science and Educational Research Council grant number GR/J16541. We would also like to thank our colleagues at CNET, particularly Jean-Bernard Stefani, Francois Horn and Laurent Hazard, for their close co-operation in this work.

## References

[1]    Coulson, G., Blair, G.S. and Robin, P., "Micro-kernel Support for Continuous Media in Distributed Systems", Computer Networks and ISDN Systems Special Issue on Multimedia, 1994; also Internal Report No. MPG-93-04 Department of Computing, Lancaster University, Lancaster LA1 4YR, UK, 1993.

[2]    Herrmann, F., Armand, F., Rozier, M., Gien, M., Abrossimov, V., Boule, I., Guillemont, M., Leonard, P., Langlois, S. and W. Neuhauser, "CHORUS, A New Technology for Building UNIX Systems", Proc. EUUG Autumn Conference, Cascais, Portugal, pp 1-18, October 3-7 1988.

[3]    Accetta, M., Baron, R., Golub, D., Rashid, R., Tevanian, A., and M. Young, "Mach: A New Kernel Foundation for UNIX Development", Technical Report Department of Computer Science, Carnegie Mellon University, August 1986.

[4]    Tanenbaum, A.S., van Renesse, R., van Staveren, H. and S.J. Mullender, "A Retrospective and Evaluation of the Amoeba Distributed Operating System", Technical Report, Vrije Universiteit, CWI, Amsterdam, 1988.

[5]    Anderson, D.P., and P. Chan, "Toolkit Support for Multiuser Audio/Video Applications", Proc. Second International Workshop on Network and Operating System Support for Digital Audio and Video, IBM ENC, Heidelberg, Germany, 1991.

[6]    Campbell, A., Coulson, G., García, F., and D. Hutchison, "A Continuous Media Transport and Orchestration Service", Proc. ACM SIGCOMM '92, Baltimore, Maryland, USA, August 1992; also ACM Computer Communication Review, Vol 22, No 4, pp 99-110, October 1992.

[7]    Tokuda, H., Nakajima, T. and Rao, P., "Real-time Mach: Towards a Predictable Real-time System", Proc. Usenix 1990 Mach Workshop, Usenix, October 1990.

[8] Reed, D.P. and Kanodia, R.K., "Synchronisation with Eventcounts and Sequences", CACM, Vol 22, No 2, pp 115-123, February 1979.

[9] Campbell, A., Coulson, G. and Hutchison, D., "A Multimedia Enhanced Transport Service in a Quality of Service Architecture", Proc. 4th International Workshop on Network and Operating System Support for Digital Audio and Video, Lancaster, UK, November 1993; also available as MPG-93-22, Computing Department, Lancaster University, Lancaster LA1 4YR, UK, 1993.

[10] Robin, P., Coulson, G., Campbell, A., Blair, G. and Papathomas, M., "Implementing a QoS Controlled ATM Based Communications System in Chorus", To be presented at IFIP Conference on High Performance Networking, Vancouver 1994; also Internal Report, MPG-94-05, Lancaster University, 1994.

[11] Liu, C.L. and Layland, J.W., "Scheduling Algorithms for Multiprogramming in a Hard Real-time Environment", Journal of the Association for Computing Machinery, Vol. 20, No, 1, pp 46-61, February 1973.

[12] Marsh, B.D., Scott, M.L., LeBlanc, T.J. and Markatos, E.P., "First class user-level threads", Proc. Symposium on Operating Systems Principles (SOSP), Asilomar Conference Center, ACM, pp 110-121, October 1991.

[13] Govindan, R., and D.P. Anderson, "Scheduling and IPC Mechanisms for Continuous Media", Thirteenth ACM Symposium on Operating Systems Principles, Asilomar Conference Center, Pacific Grove, California, USA, SIGOPS, Vol 25, pp 68-80, 1991.

[14] Shepherd, W.D., Coulson, G., García, F., and D. Hutchison, "Protocol Support for Distributed Multimedia Applications", Proc. Second International Workshop on Network and Operating Systems Support for Digital Audio and Video, Heidelberg, Germany, 1991.

[15] Clark, D.D., Lambert, M.L., and L. Zhang, "NETBLT: A High Throughput Transport Protocol", Computer Communication Review, Vol 17, No 5, pp 353-359, 1987.

[16] Tennenhouse, D.L., "Layered Multiplexing Considered Harmful", Protocols for High-Speed Networks, Elsevier Science Publishers (North-Holland), 1990.

[17] Forin, A., Golub, D. and Bershad, B., "An I/O System for Mach 3.0", Internal Report, Carnegie Mellon University, 5000 Forbes Ave., Pittsburgh, PA 15213, USA, 1990.

[18] Thekkath, C.A., Nguyen, T.D., Moy, E. and Lazowska, E., "Implementing Network Protocols at User Level", IEEE Transactions on Networking, Vol 1, No 5, pp 554-565, October 1993.

[19] Campbell, A., Coulson, G., García, F., Hutchison, D., and H. Leopold, "Integrated Quality of Service for Multimedia Communications", Proc. IEEE Infocom'93, also available as MPG-92-34, Computing Department, Lancaster University, Lancaster LA1 4YR, UK, August 1992.

[20] Cramer, A., Farber, M., McKellar, B. and Steinmetz, R., "Experiences with the Heidelberg Multimedia Communication System: Multicast, Rate Enforcement and Performance", Proc. IFIP Conference on High Speed Networks, Liege, Belgium, 1992.

[21] Coulson, G., Blair, G.S., Stefani, J.B., Horn, F. and Hazard, L., "Supporting the Real-time Requirements of Continuous Media in Open Distributed Processing", Computer Networks and ISDN Systems Special Issue on Open Distributed Processing, 1994; also Internal Report No. MPG-93-10, Department of Computing, Lancaster University, Lancaster LA1 4YR, UK, January 1993.

# 14

---

# HUMAN PERCEPTION OF
# AUDIO-VISUAL SKEW

Ralf Steinmetz

*IBM European Networking Center*
*Vangerowstraße 18 • 69115 Heidelberg • Germany*

## ABSTRACT

Multimedia synchronization comprises the definition and the establishment of temporal relationships among audio, video, and other data. The presentation of 'in sync' data streams by computers is essential to achieve a natural impression. If data is 'out of sync', the human perception tends to identify the presentation as artificial, strange, or annoying. Therefore, the goal of any multimedia system is to present all data without perceptible synchronization errors. The achievement of this goal requires a detailed knowledge of the synchronization requirements at the user interface. This paper presents the results of a series of experiments about human media perception of skew mainly related to audio and video data. It leads to a first guideline for the definition of a synchronization quality of service. The results show that a skew between related data streams may still let data appear 'in sync'. It also turned out that the notion of a synchronization error highly depends on the types of media.

## 1 INTRODUCTION

We understand multimedia according to [1,2]; a multimedia system is characterized by the integrated computer-controlled generation, manipulation, presentation, storage, and communication of independent discrete and continuous media. The digital representation of any data and the synchronization between various kinds of media and data are the key issues for integration. Multimedia synchronization is needed to ensure a temporal ordering of events in a multimedia system.

At a first glance this ordering applies to single data streams: a stream consists of consecutive logical data units (LDUs). In the case of an audio stream, LDUs may be individual samples transferred together from a source to one or more sinks. A video LDU typically corresponds to a single video frame and consecutive LDUs have to be presented at the sink with the same temporal relationship as they were captured at the source leading to intrastream synchronization.

235

The temporal ordering also applies to related data streams. The most often discussed relationship is the simultaneous playback of audio and video with 'lip synchronization'. Both kinds of media must be 'in sync', otherwise the viewer would not be satisfied with the presentation. In general an interstream synchronization involves relationships between all kind of media including pointers, graphics/images, animation, text, audio, and video. In the following, 'synchronization' always means interstream synchronization.

For delivering multimedia data correctly at the user interface, synchronization is essential. Unlike other notions of correctness, it is not possible to provide an objective measurement for synchronization. As human perception varies from person to person, only heuristic criteria can determine whether a stream presentation is correct or not. This paper presents our results of some extensive experiments related to human perception of synchronization between different media.

To reach the goal of an error-free data delivery, audio, video, and other data are often multiplexed (i.e. physically combined in one data unit) and, hence, synchronized at the source and demultiplexed just before presentation at the sink. Multiplexing is not always possible and wanted, e.g., because multimedia data needs to go through different routes in a computing system. The separate handling of previously related data leads to time lags between the media streams. These lags have to be adjusted at the sink for 'in sync' presentation.

Some work on how to implement multimedia synchronization was done in related projects [3,4,5,6,7,8]. Work has also been devoted to define synchronization requirements [9,10,11,12,13]. It is often reported that audio can be played up to 120 ms ahead of video and in the reverse situation video can be displayed 240 ms ahead of audio. Both temporal skews will sometimes be noticed, but can easily be tolerated without any inconvenience by the user [14]. Some authors report a skew of +/-16 ms [15] or no skew at all to be acceptable.

Implementing our own synchronization mechanisms, we were unable to draw the right conclusions from these reports - their statements were contradictory. There was a lack of an in-depth analysis of synchronization between the various kind of media and, in particular, for lip synchronization. We decided to conduct our own study and to explore these fundamental issues to obtain results that allow us to quantify the quality of service requirements for multimedia synchronization.

The remainder of this text is organized into six sections. Section 2 outlines the main results of lip synchronization experiments, the notion of the 'quality of synchronization' is elaborated in Section 3. Section 4 describes the test strategy, how the results were achieved including influencing factors. Remaining types of media synchronization are discussed in Section 5. Section 6 provides a comprehensive overview of all types of media skew in terms of the required quality of service parameters.

## 2 EXPERIMENTAL SET-UP

'Lip synchronization' denotes the temporal relationship between an audio and a video stream where speakers are shown while they say something. The time difference between related audio and video LDUs is known as the 'skew'. Streams which are perfectly 'in sync' have no skew, i.e., 0 ms. We conducted experiments and measured which skews were perceived as 'out of sync' for audio and video data. In our experiments, users often mentioned that something is wrong with the synchronization, but this did not disturb their feeling for the quality of the presentation. Therefore, we additionally evaluated the tolerance of the users by asking if the data out of sink affects the quality of the presentation.

In several discussions with experts working with audio and video, we noticed that most of the personal experiences were derived from very specific situations. As an immediate consequence we have been confronted with a wide range and tolerance levels up to 240 ms. A comparison and a general usage of these values is doubtful because the environments from which they resulted were incomparable. In some cases we encountered the 'head view' displayed in front of some single color background on a high resolution professional monitor. In another set-up a 'body view' was displayed in a video window at a resolution of 240*256 pixels in the middle of some dancing people. In order to get the most accurate and stringent affordable skew tolerance levels, we selected a speaker in a TV news environment as a 'talking head' (see figure 1). In this scenario, the viewer is not disturbed by background information. The user is attracted by the gestures, eyes, and lip movement of the speaker. We selected a speaker who makes use of gestures and articulates very accurately.

Figure 1    Left: Head View, Middle: Shoulder View, Right: Body View[1]

We recorded the presentation and then played it back in our experiments with artificially introduced skew that was adjusted according to the frame rate, i.e., n times 40 ms (derived from the European TV standard), which was introduced by professional video editing equipment. We conducted some experiments with a

---

1. This is just an outline of the different views, the quality of the original clips is TV-like.

higher resolution time scale by cutting the material with the help of a computer where it was possible to introduce a smaller delay in the audio stream. It turned out that there was no need for any test with higher granularity than 40 ms.

We expected a relationship between the detectable skew and the actual size of the head displayed at the monitor. As shown in Figure 1 we selected three different views of the speaker. At the very close 'head view' the head completely fills the screen, the 'shoulder view' shows the head as well as the shoulders while the third 'body view' captures the whole person sitting in a room.

Lip synchronization usually applies to speech as an acoustic signal related to its visual representation of the speaker. We expand this notion to cover the correlation between noise and its visual appearance, e.g., clapping. For the latter, our experiments included a person working with a hammer and some nails. The most exhaustive study, however, was performed in the news environment.

Figure 2[1] provides an overview of the main results. The vertical axis denotes the relative amount of test candidates who detected a synchronization error, regardless of being able to determine if audio was before or after the video. As one might expect, if the skew is relatively small most of the people did not notice it; large skews became obvious. However, our initial assumption was that the three curves related to the different views would be very different. However as shown in Figure 2 this is not the case.

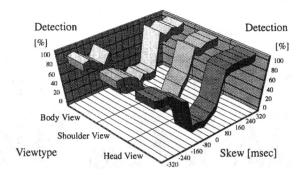

**Figure 2**   Detection of synchronization errors with respect to the three different views.

Figure 3 shows these curves in detail. A careful analysis provides us with information regarding the asymmetry, some periodic ripples and minor differences between the various views.

---

1. In all figures, negative skew denotes 'video ahead of audio', while positive skew means 'video being behind audio'.

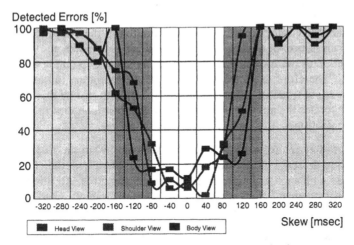

**Figure 3**     Areas related to the detection of synchronization errors

The left side of the figure relates to negative skew values, where video is ahead of audio. In our daily life, we experience this situation whenever we talk to some distant located person. All three curves are, in general, flat in this region. Since we are not accustomed to hearing speech ahead of the related visual impression, the right side of the curves turns out to be steeper.

The 'body view' curve is broader than the 'head view' curve, at the 'head view' a small skew was easier to notice. This was more difficult in the 'body view'. The 'head view' is also more asymmetric than the 'body view'. Basically, the further away we are situated, the less noticeable the error is.

At a fairly high skew, the curves show some periodic ripples. This is more obvious in the case of audio being ahead of video. It means that some people had difficulties in identifying the synchronization error even with fairly high skew values. A careful analysis of this phenomenon lead to the following explanation; At the relative minima, the speech signal was closely related to the movement of the lips which tends to be quasi periodic. Errors were easy to notice at the start, at the end, at the borders of pauses, and whenever changing drastically the mood (e.g., from an explanation style to a sudden aggressive comment). Errors were more difficult to notice in the middle of sentences. A subsequent test containing video clips with skews according to these minima (without pauses and not showing the start, the end, and changes of mood) caused problems in identifying if there was indeed a synchronization error.

The main results of about 100 test participants are captured in the different areas of figure 3:

- The **'in sync' area** spans a skew of between -80 ms (audio after video) and +80 ms (audio ahead of video). In this zone most of the users did not detect the synchronization error. Very few mention that if there is an error it does affect their notion of quality video. Additionally, we had some results where test candidates mentioned that the perfect 'in sync' clip (skew = 0ms) is 'out of sync'. Therefore, we introduced a range of uncertainty in the graph which captures these types of inconsistencies. We came to realize that lip synchronization tolerates the above mentioned skew, this result applies to any type of lip synchronization.

- The **'out of sync' areas** span beyond a skew of -160 ms and +160 ms. Nearly everybody detected these errors and was dissatisfied with the clips. Data delivered with such a skew is in general not acceptable. Additionally, often a distraction occurred; the viewer/listener became more attracted by this 'out of sync' effect than by the content itself.

- In the **'transient' area** where *audio is ahead of video*, the closer the speaker is, the easier errors are detected and reported as disturbing. The same applies to the overall resolution; the better the resolution is, the more obvious the lip synchronization errors became.

- A second **'transient' area** where *video is ahead of audio* is characterized by a similar behavior as the other transient area as long as the skew values are near the in sync area. The closer the speaker is, the more obvious the skew is. Apart from this effect we noticed that video ahead of audio can easier be tolerated than the vice versa.

This asymmetry is very plausible: In a conversation where two people are located 20 m apart, the visual impression will always be about 60 ms ahead of the acoustics due to the faster light propagation compared to the acoustic wave propagation. We are just more used to this situation than to the previous one.

Concerning the different areas, we got similar results with the noise and video experiment (hammer with nails) although the transient areas are more narrow. In this experiment, the type of view had a negligible influence. The presentation of some violinist in a concert and a choir did not show more stringent skew demands than the speaker being synchronized.

A comparison between sets of experiments ran in English and German showed no difference. There might be, however, a problem inherent with the test candidates: as Germans are used to watch synchronized films and movies, they could be less sensitive to synchronization errors than, e.g. Americans. However some minor experiments with English, Spanish, Italian, French and Swedish presented always by native speakers to native speakers verified that the specific language has almost no influence on the results.

We did not find any variation between groups of participants with different habits regarding the amount of TV and films usually watched.

## 3  QUALITY OF SKEW VALUES

Figure 3 outlines the perception of synchronization errors. More important than just to notice the error is the effect of such an 'out of sync' video clip on the human perception. If in an extreme case all people tend to like audio data to be, e.g., 40 ms ahead of video, we should take it into account. Therefore the test candidates were asked to qualify a detected synchronization error in terms of being acceptable, indifferent, or annoying. From these answers we derived the 'level of annoyance' which quantifies the quality of synchronization.

Figure 4 shows by which degree a skew was believed to be acceptable or intolerable. We used the 'talking head' experiment and depict here the 'shoulder view' as it is a compromise between the 'head' and the 'body view'.

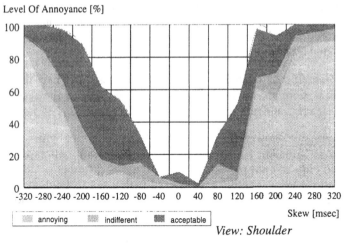

**Figure 4**   Level of annoyance at shoulder view

The dark grey areas relate to all test candidates who would accept to listen to and watch video with this synchronization error. In a small follow-on experiment we selected a few test candidates who would tolerate such a skew and presented them a whole movie with this synchronization error. We chose a skew of -160 ms (video ahead of audio). They did not complain and very soon concentrated on the content instead of being attracted by looking for some synchronization problem. The curve at the bottom of the dark grey area shows an obvious asymmetry which occurs due to the more natural acceptance of visual perception being ahead of related acoustic impression.

The light grey area relates to all people who really dislike this skew and were distracted by it. It also contains the asymmetry discussed above. During the evaluation phase of this study on synchronization, we introduced a skew of +80 ms and -80 ms into two whole movies. These movies were shown to a few candidates who mentioned that such a skew is annoying. It turned out after a short discussion that, if we introduced this artificially (or if we cheated), they did not object at all. The same experiment with a skew of -240 ms or +160 ms would lead to a real distraction from the content and to a severe feeling of annoyance.

This evaluation of the level of annoyance provides a further argument for allowing the skew of lip synchronization to take values between -80 ms and +80 ms as mentioned in the former section.

## 4  TEST STRATEGY

For each person, the lip synchronization test took approximately 45 minutes. The experiment was intentionally carried out with the same audio and video over and over again. This led to some concentration problems during the whole test, which were alleviated by introducing breaks.

We always ran all tests related to one view in one session. Then, the second and the third view were shown in their sessions. The order of the sessions had no effect. Individual samples, each having a different skew, were shown randomly.

Initial experiments showed that a total length of about 30s with a small subsequent break is sufficient for getting the users impression. All experiments with longer video clips did not provide any additional or different results. With some test candidates, who were more experienced with video technology and synchronization issues, 5s turned out to be sufficient. Nevertheless we decided to use 30s for each sample.

The background of all scenes was static (i.e., not moving) and out of focus in order to keep the distraction to a minimum. We focussed on the detection of synchronization errors in the most challenging set-ups, this allowed the determination of skew values independently from the actual content of the video and audio data. In these experiments the viewer should never have been distracted by the background.

The same consideration, i.e. background vs. foreground, can be applied to the audio data. The voice of the speaker can be mixed with some background noise or music. In order to differentiate between foreground and background, the volume of the speaker should be at least twice the volume of the background audio.

In contrast to the video analogy discussed in the previous paragraph, any background audio did not influence our results.

The group of people was selected according to an equal distribution of sex and ages. To have a representative distribution we did not take into account habits (like the time spent for watching TV) and the social status or any other characteristics of the test candidates.

It would have been very interesting if - before presenting each sample - the candidates were not aware of the fact that we were looking for synchronization issues. As soon as the test candidates noticed the first time a synchronization fault, they would not have been allowed to continue the experiment with further skews. This would have led to results for casual unexperienced users. As a matter of fact, we started to run the experiment in this way with very few people. It turned out that lip synchronization is not detected so easily leading to a broader range of the 'in sync' zone.

In order to provide results for building multimedia systems for all types of users, we have to make the assumption that a user can also make frequent use of such a system and interact for a longer time with the application. Therefore, the results of users being aware of possible synchronization faults provide the correct basis.

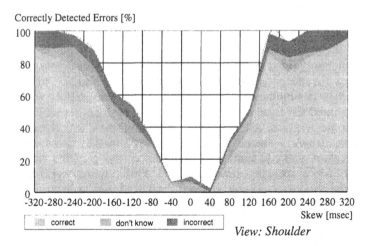

**Figure 5** Correct detection of the perceived synchronization error

For the purpose of double checking, the candidates were asked to define exactly which type of synchronization error they noticed. It is easier just to detect that something is wrong than to precisely state if audio is ahead of video or vice versa. Figure 5 summarizes the results of correct perception of the skew in the 'shoulder view' scenario.

Near the error-free synchronization value (at 0 ms) it was difficult to determine the type of skew, as soon as the values ranged beyond -40 ms or +40 ms almost everybody provided correct answers.

# 5 MEDIA DEPENDENCY OF SKEW

Lip synchronization was investigated by us because of the contradictory results available from other sources. In the following, we will summarize other synchronization results which we found provided less diverse statements. We do this to arrive at a more complete picture of synchronization requirements.

Since the beginnings of digital *audio* the ('affordable') *jitter* and the jitter to be tolerated by dedicated hardware has been studied. In discussions with Dannenberg, he provided us some references and the following explanations of these studies: In [17] the maximum allowable jitter in a sample period for at 16 bit quality audio is mentioned to be 200ps, this is explained as the error equivalence to the magnitude of the LSB (least significant bit) of a full-level maximum-frequency 20KHz signal. In [18] some perception experiments recommend an allowable jitter in an audio sample period between 5 and 10 ns. Further perception experiments were carried out by [19] and [20], the maximum spacing of short clicks to obtain fusion into a single percept was mentioned to be 2ms (as cited by [14]).

In a computer-supported cooperative work (CSCW) environment, cameras and microphones are attached to the users' workstations. The audio and video data of one participant is simultaneously presented at the other remote workstation(s), e.g., we assumed the issue of the discussion is a business report including data related to some graphics. All participants have a window with these graphics on their desktop where a shared telepointer is used in the discussion. With the pointer, speakers point to individual elements of the graphics which they are referring to while speaking simultaneously. This requires synchronization of *audio and the remote telepointer*. Using the same margins as in our lip synchronization experiments, the 'in sync' area related to audio ahead of pointing is 750 ms and for pointing ahead of audio it is 500 ms [25]. In most of the daily occurring discussions using a telepointer these results can be relaxed. This zone allows for a clear definition of the 'in sync' behavior regardless of the content.

The combination of *audio and animation* is usually not as stringent as lip synchronization. A multimedia course on dancing, e.g., comprises the dancing steps as animation with the respective music. By making use of the interactive capabilities, individual sequences can be viewed and listened to over and over again. In this example the synchronization between music and animation is particularly important. Experience shows that a skew of +/- 80 ms fulfills all user demands even though some jitter may occur. Nevertheless, the most challenging issue is the correlation between a noisy event and its visual representation, e.g. a simulated crash of cars. Here we encounter the same constraints as for lip synchronization, +/- 80 ms.

*Two audio tracks* can be tightly or loosely coupled, the effect of related audio streams depends heavily on the content:

- A stereo signal usually contains information about the location of the sources of audio and is *tightly coupled*. The correct processing of this information by the human brain can only be accomplished if the phases of the acoustic signals are delivered correctly. This demands for a skew less than the distance between consecutive samples leading to the order of magnitude of 20 μs. [22] reports that the perceptible phase shift between two audio channels is 17μs. This is based on a headphone listening experiment. Since a varying delay in one channel causes the sound's source location apparently to move, Dannenberg proposed to allow an audio sample skew between stereo channels within the boundaries of +/- 11μs. This is derived from the observation that a one-sample offset at a sample rate of 44kHz can be heard.

- *Loosely coupled* audio channels are a speaker and, e.g., some background music. In such scenarios we experience an affordable skew of 500 ms. The most stringent loosely coupled configuration has been the playback of a dialogue where the audio data of the participants originate from different sources. The experienced acceptable skew was 120 ms.

The combination of *audio with images* has its initial application in slide shows. By intuition a skew of about 1s arises which can be explained as follows [15]: Consider that it takes a second or so to advance a slide projector. People sometimes comment on the time it takes to change transparencies on an overhead projector, but rarely worry about automatic slide projectors.

A more elaborated analysis leads to the time constraints equivalent to those of pointer synchronization. The affordable skew decreases as soon as we encounter *music* played in correlation *with notes* for, e.g., tutoring purposes. [15] points out that here an accuracy of 5 ms is required: Current practice in music synthesizers allows delays ranging up to 5 ms, but jitter is less than total delay. A 2 ms number refers to the synchronization between the onset times of two nominally simultaneous notes or the timing accuracy of notes in sequence, see also [23,21,24].

The synchronized presentation of *audio with some text* is usually known as audio annotation in documents or, e.g., part of an acoustic encyclopedia. In an existing 'music dictionary', an antique instrument is described and is played simultaneously. An example for a stronger correlation is the playback of a historical speech of, e.g. J.F. Kennedy with simultaneous translation into German text. This text is displayed in a separate window and must relate closely to the actual acoustic signals. Karaoke systems are another good example of necessary audio and text synchronization.

For this type of media synchronization the affordable skew can be derived from the duration of the pronunciation of short words which last approximately 500 ms. Therefore the experimentally verified skew of 240 ms is acceptable

The synchronization of *video and text* or *video and image* occurs in two different fashions:

- In the *overlay mode*, the text is often an additional description to the displayed moving image sequence. In a video of playing billiard, the image is used to denote the exact way of the ball after the last stroke. The simultaneous presentation of the video and the overlaid image is important for the correct human perception of this synchronized data. The same applies to a text which is displayed in conjunction with the related video images: Instead of having the subtitles always located at the bottom, it is possible to place text close to the respective topic of discussion. This would cause an additional editing effort at the production phase and may not be for the general use of all types of movies but, for tutoring purposes some short text near by the topic of discussion is very useful. In such overlay schemes, this text must be synchronized to the video in order to assure that it is placed at the correct position. The accurate skew value can be derived from the minimal required time. A single word should appear on the screen in order to be perceived by the viewer: 1 s is certainly such a limit. If the media producer wants to make use of the flash effect, then such a word should be on the screen for at least 500 ms. Therefore, regardless of the content of the video data we encounter 240 ms to be absolutely sufficient.

- In the second mode *no overlay* occurs, skew is less serious. Imagine some architectural drawings of medieval houses being displayed in correlation with a video of these building: While the video is showing today's appearance, the image presents the floor plan in a separate window. The human perception of even simple images requires at least 1 s, we can verify this value with an experiment with slides: the successive projector of non-correlated images requires about 1 s, as the interval between the display of a slide and the next one in order to catch some of the essential visual information of the slide. A synchronization with a skew of 500 ms (half of this mentioned 1 s value) between the video and the image or the video and text is sufficient for this type of application.

Sometimes *video* is combined *with animation* as there may be a film where some actors become animated pictures. But, for the following short reasoning of synchronization between video and animation let us go back to the example of a video showing the stroke of a billiard ball and the image of the actual 'route' of this ball. Instead of the static image, the track of the ball can be followed by an animation which displays this route at the time the ball is moving on the table. In this example any 'out of sync' effect is immediately visible. In order for humans to be able to watch the ball with the perception of a moving picture, this ball must be visible in several consecutive adjacent video frames at a slightly different position: a good result can be achieved, if in every 3 subsequent video frames, the ball moves by the distance of it's diameter. Less frames will result in the problem of visibility of what occurs, e.g., in tennis, and it may lead to difficulties with the notion of continuity. Derived from this number of 3 subsequent frames, we allow the equivalent skew of 120 ms to occur. This is very tight synchronization, and we have not found any practical requirement which cannot be handled with this value of the affordable skew.

Multimedia systems also incorporate the real-time processing of *control data* and the presentation of this data using various media. A tight timing requirement occurs if the person has to react to this displayed data. No overall timing demand can be stated as these issues highly depend on the application itself.

# 6 QUALITY OF SERVICE

The control of synchronization in distributed multimedia systems requires a knowledge of the temporal relationship between media streams. The result of this study is of service to this management component. Synchronization requirements can be expressed by a quality of service (QoS) definition. This QoS parameter defines the acceptable skew within the involved data streams, it defines the affordable synchronization boundaries. The notion of QoS is well established in communication systems; in the context of multimedia, it also applies to local systems. If the video data is to be presented simultaneously to some audio and both are stored as different files or as different entries in a database, lip synchronization according to the above mentioned results has to be taken into account.

In this context we want to introduce the notion of *presentation* and *production level synchronization*:

- *Production level synchronization* refers to the QoS to be guaranteed prior to the presentation of the data at the user interface. It typically involves the recording of synchronized data for a subsequent playback. The stored data should be captured and recorded with no skew at all, i.e. totally "in sync". This is of particular interest if the file is stored in an interleaved format applying multiplexing techniques. Imagine a participant of an audio-video conference who additionally records this audiovisual data to be played back later for a remote spectator. At the conference participant's site, the actual incoming audiovisual data is 'in sync' according to the defined lip synchronization boundaries. Let the data arrive with a skew of +80 ms and let audio and video LDUs be transmitted as a single multiplexed stream over the same transport connection. It is displayed to the user and directly stored on the harddisk (still having this skew). Later on, this data is presented simultaneously at a local workstation and to the remote spectator. For a correct data to be deliverable, the QoS should be specified as being between -160 ms and 0 ms. At the remote viewer's station - without this additional knowledge of the actual skew - it might turn out that by applying these boundaries twice, data is not 'in sync'. In general, any synchronized data which will be further processed should be synchronized according to a production level quality, i.e. with no skew at all.

■ The whole set of experiments discussed in this report identifies *presentation level synchronization*, it focuses on the human perception of synchronization and defines whatever is reasonable at the user interface. As shown in the above paragraph, by recording the actual skew as part of the control information, the required QoS for synchronization can be easily computed. Therefore, in advanced systems, data may also be recorded 'out of sync' leading to an 'in sync' presentation.

The required QoS for synchronization is expressed as the allowed skew. The QoS values shown in Table 1 relate to presentation level synchronization. Most of them result from exhaustive experiments and experiences, others are derived from the literature as referenced in the paper. To our understanding, they serve as a general guideline for any QoS specification. During the lip synchronization experiment we learned that there are many factors such as the distance of a speaker which to some extend influence these result. We understand that this whole set of QoS parameters as first order result to serve as a general guidance. These values may be relaxed using the knowledge on the actual content.

# 7 OUTLOOK

Synchronization QoS parameters allow the builders of distributed multimedia and communication systems to make use of the affordable tolerances. In our Heidelberg multimedia system, the HeiRAT component [26] is in charge of the resource management. HeiRAT accepts QoS requests from the applications and serves for this QoS demands as interface to the whole distributed system. It makes use of the flow specification of the ST-II multimedia internetwork protocol to negotiate them among the whole set of involved system components [27]. It provides a QoS calculation by optimizing one QoS parameter dependent on the

| Media | | Mode, Application | QoS |
|---|---|---|---|
| video | animation | correlated | +/- 120 ms |
| | audio | lip synchronization | +/- 80 ms |
| | image | overlay | +/- 240 ms |
| | | non overlay | +/-500 ms |
| | text | overlay | +/- 240 ms |
| | | non overlay | +/-500 ms |
| audio | animation | event correlation (e.g. dancing) | +/- 80 ms |
| | audio | tightly coupled (stereo) | +/- 11 µs |
| | | loosely coupled (dialogue mode with various participants) | +/- 120 ms |
| | | loosely coupled (e.g. background music) | +/- 500 ms |
| | image | tightly coupled (e.g. music with notes) | +/- 5 ms |
| | | loosely coupled (e.g. slide show) | +/- 500 ms |
| | text | text annotation | +/- 240 ms |
| | pointer | audio relates to showed item | -500 ms, + 750 ms[a] |

**Table 1**   Quality of Service for synchronization purposes

a.pointer ahead of audio for 500 ms, pointer behind audio for 750 ms [25]

resource characteristics. Subsequently resources are reserved according to the QoS guarantees. At the actual data transfer phase, resources are scheduled according to the provided guarantees.

Synchronization is a crucial issue of multimedia systems. In local systems it is often easy to provide because there are sufficient resources or it is a single user configuration. In networked systems we encounter a plethora of concurrent processes making use of the same scarce resources. A skew between media easily arises.

This paper provides a set of quality of service values for synchronization. It is a feather in our cap to reach results for wide range of media synchronization with extensive user interface experiments. As the next step, we currently run experiments related to the affordable jitter in continuous media presentations. The enforcement of which remains to be a different item which already has been addressed by several systems. The presentation of audio and video, according to some logical time system is one of the possible solutions.

## 8 ACKNOWLEDGMENTS

First of all I would like to acknowledge the enthusiastic work done by Clemens Engler: We spent hours and nights of controversial discussions on the expected results, the influencing factors and the design of the experiments. Clemens Engler also carried out most of the experimental work. Wieland Holfelder helped in producing the basic video material, and I would like to acknowledge the patience and accuracy of all our test candidates. Roger Dannenberg, CMU Pittsburgh, provided many valuable hints concerning jitter of audio samples and synchronization related to music. Ralf Guido Herrtwich substantially commented the final version of the paper. Thank you.

## REFERENCES

[1] Ralf Steinmetz, Ralf Guido Herrtwich: "Integrated Distributed Multimedia-Systems," Informatik Spektrum, Springer Verlag, vol.14, no.5, October 1991, pp. 280-282.

[2] Ralf Steinmetz, "Multimedia-Technology: Fundamentals (in German: "Multimedia Technologie: Einführung und Grundlagen")," Springer-Verlag, September 1993.

[3] David P. Anderson, George Homsy: "Synchronization Policies and Mechanisms in a Continuous Media I/O Server," International Computer Science Institute, Technical Report no. 91-003, Berkeley, 1991.

[4] Gerold Blakowski, Jens Huebel, Ulrike Langrehr, Max Muhlhaeuser: "Tools Support for the Synchronization and Presentation of Distributed Multimedia," computer communications, vol. 15, no. 10, December 1992.

[5] L.Li, A. Karmouch, N.D. Georganas, "Synchonization in Real Time Multimedia Data Delivery," IEEE ICC'92, Chicago, USA, June 1992. pp. 322.1.

[6] L.Li, L. Lamont, A. Karmouch, N.D. Georganas: "A Distributed Synchronization Control Scheme in A Group-oriented Conferencing Systems," Proceedings of the second international conference, Broadband Islands, Athens, Greece, June 15-16, 1993

[7] Doug Shepherd, Michael Salmony: "Extending OSI to Support Synchronisation Required by Multimedia Applications," Computer Communications, vol.13, no.7, September 1990, pp. 399-406.

[8] Ralf Steinmetz: "Multimedia Synchronization Techniques: Experiences Based on Different System Structures," IEEE Multimedia Workshop '92, Monterey, April 1992.

[9] Thomas D.C. Little, Arif Ghafoor: "Network Considerations for Distributed Multimedia Objects Composition and Communication," IEEE Network Magazine, vol. 4 no. 6, November 1990, pp. 32-49.

[10] Thomas D.C. Little, A. Ghafoor: "Synchronization and Storage Models for Multimedia Objects," IEEE Journal on Selected Areas in Communication, vol. 8, no. 3, Apr. 1990, pp. 413-427.

[11] Cosmos Nicolaou: "An Architecture for Real-Time Multimedia Communication Systems," IEEE Journal on Selected Areas in Communication, vol. 8, no. 3, April 1990, pp. 391-400.

[12] Kaliappa Ravindran: "Real-time Synchronization of Multimedia Datastreams in High Speed Packet switching Networks," in Workshop on Multimedia Information Systems (MMIS '92), IEEE Communications Society, Tempe, AZ, February 1992.

[13] Ralf Steinmetz: "Synchronization Properties in Multimedia Systems," IEEE Journal on Selected Areas in Communication, vol. 8, no. 3, April 1990, pp. 401-412.

[14] Alan Murphy: "Lip Synchronization," personal communication on a concerning set of experiments, 1990.

[15] Roger Dannenberg: Personal communication on sound effects and video synchronization and on music play back and visualization of the corresponding strokes, 1993.

[16] Harald Rau, personal communication, 1993.

[17] Barry Blesser: "Digitization of Audio: A Comprehensive Examination of Theory, Implementation, and Current Practice, Journal of the Audio Engineering Society, JAES 26(10), October 1978, pp. 739-771.

[18] T. Stockham: "A/D and D/A Converters: Their Effect on Digital Audio Fidelity," in Digital Signal Processing, L. Rabiner and C. Rader, (Eds.), IEEE Press, NY 1972.

[19] J.C.R. Licklider: "Basic correlates of the auditory stimulus," in S. S. Stevens, ed. Handbook of Experimental Psychology, Wiley, 1951.

[20] H. Woodrow: "Time Perception," in S. S. Stevens, (Ed.) Handbook of Experimental Psychology, Wiley, 1951.

[21] Dean Rubine, Paul McAvinney: "Programmable Finger-tracking Instrument Controllers," Computer Music Journal, vol. 14, no. 1, Spring 1980, pp. 26-41.

[22] Roger Dannenberg, Richard Stern: Experiments Concerning the Allowable Skew of Two Audio Channels Operating in the Stereo Mode, personal communication, 1993.

[23] M. Clynes: "Secrets of Life in Music: Musicality Realized by Computer," in Proceedings of the 1984 International Computer Music Conference, San Francisco, International Computer Music Association, 1985.

[24] M. Stewart: "The Feel Factor: Music with Soul," Electronic Musician, vol. 3, no. 10, pp. 55-66, 1987.

[25] Ralf Steinmetz, Clemens Engler: "Human Perception of Media Synchronization," IBM Technical Report 43.9310, IBM European Networking Center, Heidelberg, 1993.

[26] Carsten Vogt, Ralf Guido Herrtwich, Ramesh Nagarajan: "HeiRat: The Heidelberg Resource Administration Technique Design Philosophy and Goals," IBM Technical Report 43.9307, IBM European Networking Center, Heidelberg 1992.

[27] Luca Delgrossi, Ralf Guido Herrtwich, Frank Oliver Hoffmann: "An Implementation of ST-II for the Heidelberg Transport System," IBM Technical Report 43.9303, IBM European Networking Center, Heidelberg, 1993

# 15

# CINEMA - AN ARCHITECTURE FOR DISTRIBUTED MULTIMEDIA APPLICATIONS

K. Rothermel, I. Barth, T. Helbig

*University of Stuttgart,*
*Institute of Parallel and Distributed High-Performance Systems,*
*Breitwiesenstraße 20-22, D-70565 Stuttgart, Germany*

## ABSTRACT

Distributed multimedia applications combine the advantage of distributed computing with the capability of processing discrete and continuous media in an integrated fashion. The development of multimedia applications in distributed environments requires specific abstractions and services, which are usually not provided by generic operating systems. These services are typically realized by software components, often referred to as middleware.

The CINEMA (Configurable INtEgrated Multimedia Architecture) project aims at the development of powerful abstractions for multimedia processing in distributed environments. This paper presents a flexible mechanism for the dynamic configuration of applications. The proposed mechanism allows for the definition of arbitrary complex flow graphs connecting various types of multimedia processing elements. Further, processing elements can simply be composed from other ones to provide higher levels of abstraction. We also propose the abstraction of a clock hierarchy to permit grouping, controlling, and synchronization of media streams.

## 1 INTRODUCTION

Advances in the computer and communication technology have stimulated the integration of digital audio and video with computing, leading to the development of distributed multimedia systems. This class of systems combines the advantages of distributed computing with the capability of processing discrete media, such as text or images, and continuous media, such as audio or video, in an integrated fashion. The capability of integrated multimedia processing not only enhances conventional application environments, but also opens the door for new and innovative applications. A major advantage of multimedia computing in distributed environments is the possibility of sharing resources among applications and users where shared resources may be data objects such as multimedia titles, special processing elements such as compression modules, or special devices such as professional VCRs.

The processing and communication of media streams requires specific system services. In general, media streams are associated with a certain quality that has to be maintained by the underlying system. To be able to guarantee the required stream quality, system services for allocating and reserving system resources, such as CPU cycles or network bandwidth, are needed. Moreover, applications need to control the flow of streams, i.e. they should be able to start, pause, continue or scale individual streams. In many scenarios, it is desirable to group related streams and to control groups of streams rather than individual streams. Finally, powerful services to synchronize multiple streams are required. Those services should permit applications to specify which streams are to be synchronized and how these streams temporally relate to each other.

Generic operating systems usually do not provide those specific multimedia services. The gap between the functionality offered by operating systems and the specific needs of distributed multimedia applications is closed by software components often referred to as middleware. The CINEMA (Configurable INtEgrated Multimedia Architecture) system, which is currently under development at the University of Stuttgart, belongs to this system category. It provides abstractions for the dynamic configuration of distributed multimedia applications. Clients may define arbitrary data flow graphs, connecting various processing elements called components. Moreover, component nesting is supported to achieve higher levels of abstractions by simply composing more complex components from already existing ones. The abstraction of a session allows for atomic resource allocation and reservation for any group of connected components. CINEMA provides the concept of a clock hierarchy for grouping and controlling streams and groups of streams. The same abstraction permits to express arbitrary complex stream synchronization requirements.

The remainder of the paper is organized as follows. In the next section, a brief overview of related work is given. Then, in section 3, the way how applications are configured in CINEMA is described in some detail. This section also introduces the concept of component nesting. The abstractions for grouping, controlling and synchronizing media streams and groups of streams are presented in section 3.1. Finally, we conclude with a brief summary.

## 2  RELATED WORK

The multitude of problems that arise when integrating multimedia processing into conventional computer systems and attempting to develop distributed multimedia applications are addressed in several projects, which put emphasis on different issues. In the SUMO project [1], the Chorus [2] micro-kernel is extended to support continuous media. This is done by using the real-time features of Chorus and adding stream-based data transfer and quality of service control inside the operating system. The features are accessible by a low-level API. The focus of this work is on operating system issues like scheduling, but not on providing a universal platform and high-level abstractions for developing and configuring distributed multimedia applications. The problem of configuring distributed applications by using software components that are interconnected by linked ports is addressed by Conic [3] and its follow-up project REX [4]. Conic offers languages for programming components and configuring applications without supporting multimedia data handling. The configuration process is centralized in a configuration manager which accepts change specifications for altering configurations.

Specific abstractions for controlling multimedia data streams have been proposed as well. Some of them apply to non-distributed environments only (e.g. Quick-Time [5] or IBM's Multimedia Presentation Manager [6]), while others are tailored to specific configurations (e.g. ACME [7] and Tactus [8]), and mainly are extensions of network window systems supporting streams of digital audio and video data. General requirements that should be met by architectures supporting distributed multimedia applications are specified in the Request for Technology [9] of the Interactive Multimedia Association (IMA). A response to this request contributed by some companies [10] proposes abstractions to structure and control distributed multimedia environments while using multi-vendor processing equipment. The proposal assumes generic multimedia processing elements producing and consuming multimedia data via ports that are associated with formats. However, the nesting of processing elements is not supported and, although grouping is used to handle resource acquisition, stream control and specification of end-to-end quality of service, no means to specify synchronization relationships between data streams are provided.

# 3 CONFIGURATION OF MULTIMEDIA APPLICATIONS

In order to build large software systems, it is necessary to decompose a system into modules each of which can be separately programmed and tested. The system is then composed as a configuration of these software components. Component programming and component configuration are separate activities which have been referred to as "programming-in-the-small" and "programming-in-the-large", respectively [11].

Configuration may be static or dynamic. In the first approach to system building, all components of the system are configured at the same time. If a modification of the system is required, the complete system has to be stopped and rebuilt according to the new configuration specification. Obviously, static configuration is not a feasible approach in the context of distributed multimedia systems, in which configurations often depend on the available resources and the quality of service the user asks for at run time. Moreover, multimedia applications are often highly dynamic in the sense that users may join and leave the application during run time. Usually, each change in the user community implies a modification of the configuration. Examples for these applications can be found in the area of video conference systems or CSCW systems. Consequently, for multimedia systems the ability to extend and modify a system while it is running definitely is required. The approach of dynamic configuration provides this ability: new components can be introduced, existing ones may be replaced and the interconnection of components can be modified at run time.

In *CINEMA*, an application consists of at least one client and a set of data flow graphs. In a data flow graph, the nodes represent components, while the edges are communication links interconnecting the components. A component provides the basic abstractions for the processing of continuous media streams, such as video or audio streams. A continuous media stream is defined to be a sequence of data units, each of which is associated with a media time (for a detailed definition e.g. see [12]). The nature of a component's processing depends on the type of the component. We distinguish between source components, which produce (e.g. capture) data streams, sink components that consume (e.g. play-out) streams, and intermediate components acting as both consumers and producers (e.g. filters or mixers). Media streams may originate at multiple sources, traverse a number of intermediate components and end at multiple sinks.

A client is a software entity that - by using the *CINEMA* services - defines data flow graphs and controls the flow of data within these graphs during run time. It configures (its portion of) an application just by naming the components to be used and interconnecting them according to the application logic that has to be achieved. Furthermore, it may dynamically change the initial configuration during run time as needed. A data flow graph may be arbitrarily distributed over several nodes of a distributed system. As will be seen below, components are configuration independent, which means that their internal logic is independent of the configuration they are used in. Thus, from the client's point of view, there is no conceptual difference whether two adjacent components run either on the same node interconnected by a local link or on different nodes connected by a remote link.

A client may only control the flow of streams in the flow graphs defined by itself. In particular, a client may start, halt or scale data streams only in its so-called **application domain**, which is defined to be the set of data flow graphs specified by this client. Depending on the type of application, one or more clients may par-

**Figure 1**    Application Domains in a Conferencing Scenario

ticipate in the process of configuring the application. If multiple clients participate, the application is structured into several application domains, one for each
participant. Each client only knows and controls the objects in its domain. When
sharing components between clients, their domains overlap. The overlapping
portions contain the shared components. In other words, shared components may
be controlled by multiple clients. Refer to the simple conferencing scenario
depicted in Figure 1 for an example. In this scenario, the application consists of
several domains, each of which links two components - a virtual microphone and
speaker of a given user - to a shared mixer component. Whenever a new user
joins the application, a new domain linking the new user's (virtual) microphone
and speaker to the shared mixer is added.

After this brief overview of the process of configuration in CINEMA, we can now
take a closer look at the concepts provided for defining flow graphs, which are
components, ports and links.

## 3.1  Components and Ports

The processing of continuous media data streams is done by software and hardware modules, called devices. Devices may be e.g. microphones or speakers having specific hardware interfaces and software drivers. In CINEMA, the processing
functionality is abstracted by components. When creating a component, a client
specifies the devices that are to be used. Components consume data units of
streams reading from their input ports and produce data by writing to their output
ports. To build up data flow graphs, components are interconnected by links
between the components' ports.

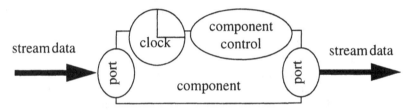

**Figure 2**    The Component's Interfaces

From the client's point of view, a component offers different interfaces to control and manipulate its behavior, the component control interface, the clock interface, and the port interface. The **component control interface** is used to access state information of a component and alter its stream handling behavior. It is specific in the sense that it depends on the processing function performed by the component. For example, the interface of a component abstracting from a speaker device may provide a method to adjust the volume of the presentation. The **clock interface** is optional for sources and mandatory for sinks and is used to control the flow of data units. A detailed description of clocks is given in Section 4. The **port interface** is used by components to send stream data to other components that are interconnected by links or to receive data from them. This decouples the multimedia processing from the transmission of data units between processing stages and allows the usage of the same component in scenarios having local as well as remote communication. To be able to check mismatching connections, each port is associated with a stream type. If a component handles multiple stream types, a new stream type containing the others may be defined. In Figure 3, we show an example of a stream type hierarchy. In this example, a port of type "video" can be connected to either one of type "video", "video-grey", or "video-color". In a stream type hierarchy, the descendents of a node are specializations of this node.

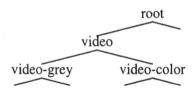

**Figure 3**   Stream Type Hierarchy

The interfaces described above are used by clients to control components and to connect them to build up data flow graphs. In the *CINEMA* system, components are managed by additional interfaces. An example for such an internal interface is the resource allocation interface, that is used to negotiate the quality of service and to reserve the required resources to ensure it.

After looking at the interfaces provided by components, we now focus on the definition of components. Configuration independence [3] is a major property to build up components that can be used in a dynamically configured distributed system. This makes it possible to use a component in arbitrary configurations without having to change its processing functionality. Configuration independence is achieved by developing components using a special programming language and compiling and linking them to independent objects. In *CINEMA*, we use an object-oriented programming language, the Component Programming Language (CPL) that is based on C++, to program components. It allows the creation of a class hierarchy with inheritance to build up specialized component classes out of existing ones. The following example shows the programming of a microphone component in CPL:

```
COMPONENT microphone
    :: SOURCE // class to derive from
    // define method to map devices
  MAP ( device MICRO ); // device parameter
    dev_name = MICRO;    // handle device parameter
  ENDMAP
    // define method to initialize component
  INIT (int sensitivity);           // specific client-IF
    dev = open(dev_name,"r");       // open the device
    dev_set_samplerate(dev,8000); // rate = 8000 Hz
    dev_set_sensitivity(dev,sensitivity); //set value
  ENDINIT
    // stream type definition
  TYPE 8kHz_Audio :: Audio; // derive 8KHz_Audio from Audio
    // definition of port named audio
  OUTPORT audio 8kHz_Audio;
    // define method to adjust microphone's sensitivity
  METHOD int sensitivity_adjust(int sensitivity)
    result = dev_get_sensitivity(dev);      // get value
    dev_set_sensitivity(dev, sensitivity); // set value
    return result; // return old value
  ENDMETHOD
    // definition of stream-handling function
  ACTION
    data = dev_get_data(dev); // get audio samples
    audio->put(data); // put samples to output port
  ENDACTION
ENDCOMPONENT
```

In the *CINEMA* system, the code segments of a component are executed in different threads. The stream handling segment, defined in the `ACTION` clause, is periodically executed in a real-time thread, whereas the methods of the component control interface are executed in a non-real-time thread. Resource requirements of the real-time thread are calculated when a session, which in *CINEMA* is the abstraction for atomic resource reservation, is established (see Section 4.1).

## 3.2  Creation of Data Flow Graphs

So far, we have introduced the definition of components, the functional building blocks. In this section, we will describe how a client builds up data flow graphs by connecting the components' ports by means of links.

To build an application, a client first establishes the processing functionality by creating the appropriate components. This is done by using a library with a set of functions and classes that is provided by the *CINEMA* system. No specialized configuration language is needed which offers the advantage to expand and shrink applications dynamically at run time depending on actual requirements. Moreover, it allows the integration of multimedia processing functionality into existing (non-multimedia) applications. Creating and accessing components does not differ from accessing normal C++-objects. It is done by using appropriate object methods.

As shown above, components may be shared by multiple clients if more than one client participates in the configuration of an application. In *CINEMA*, shared components are associated with a globally unique identifier. All clients sharing a given component create this component in their application domain by providing the component's global identifier. Of course, only the create operation issued first establishes the component, while all succeeding ones just enable the callers to access the (already existing) component.

The following code fragment shows the creation of the component objects in the conferencing example illustrated in Figure 1. The mixer component is defined as a shared component using the global identifier `conference`.

```
micro   = COMPONENT("microphone",micro_dev);
mixer   = COMPONENT("audio_mixer",NULL,"conference");
speaker = COMPONENT("speaker",speaker_dev);
```

For component initialization, each component provides a method called `init`. The code example below initializes the microphone and the speaker component and specifies the sensitivity to 50 and the volume to 40. The initialization has to be done before defining a session.

```
micro   ->init("sensitivity",50);
speaker->init("volume",40);
```

After component objects have been created, they are connected by creating links among their ports. The component's port objects are accessed by using the method `port` in connection with the port identifier. In our code fragment, we link the output port of the microphone component (named `audio`) and the input port of the mixer component (named `audio_in`). A second link is established between the output port of the mixer component (`audio_out`) and the input port of the speaker component (`audio`).

```
link(micro->port("audio"), mixer->port("audio_in"));
link(mixer->port("audio_out"), speaker->port("audio"));
```

It is important to mention that building up a data flow graph only describes the topology of an application. Linking components does not imply the reservation of resources. To enable communication, sessions have to be established.

## 3.3  Nesting Components

In many areas, nesting has turned out to be a very powerful concept for building higher levels of abstractions. In *CINEMA*, more complex components, called **compound components**, can be composed from other components. Compound components contain a part of a data flow graph. They are used like non-nested, basic components, i.e. from the client's point of view, there is no difference in using basic or compound components since the internal structure of compound components is hidden.

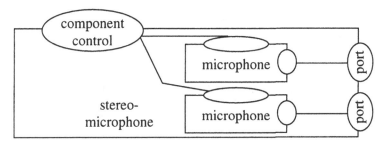

**Figure 4**   Compound Component

Constructing compound components from existing ones is straightforward. Instead of programming an ACTION clause, a part of a data flow graph is defined using already existing components. The components used to build the compound component are declared in the USE clause. The way they are interconnected by links is defined in the LINK clause. Component control interfaces of the nested components are accessed through a common interface provided by the compound component. The mapping of these interfaces is defined when building a compound component.

As an example for the programming of a compound component (see Figure 4), we show the definition of a component representing a stereo microphone component. This component uses two components of class microphone as they were declared in Section 3.1.

```
COMPONENT stereo_micro
    // define Method to map devices
  MAP ( device MICRO_l, device MICRO_r );
    dev_MICRO_l = MICRO_l; // handle device parameter
    dev_MICRO_r = MICRO_r; // handle device parameter
  ENDMAP
    // define mothod to initialize component
  INIT ( int sensitivity );
    // init nested components with provided parameters
```

```
    micro_l->init(sensitivity); // initialize micro_l
    micro_r->init(sensitivity); // initialize micro_r
  ENDINIT
    // define the ports of the compound component
  OUTPORT audio_l 8kHz_Audio;
  OUTPORT audio_r 8kHz_Audio;
    // create component objects
  USE
    micro_l = COMPONENT("microphone",dev_MICRO_l);
    micro_r = COMPONENT("microphone",dev_MICRO_r);
  ENDUSE
    // build up flow graph with links
  LINK   // use "this" to refer to compound component
    link(micro_l->port("audio"),this->port("audio_l"));
    link(micro_r->port("audio"),this->port("audio_r"));
  ENDLINK
    // map the specific interfaces to nested ones
  METHOD int sensitivity_adjust(int sensitivity)
    result = micro_l->sensitivity_adjust(sensitivity);
    result = micro_r->sensitivity_adjust(sensitivity);
    return result; // return value
  ENDMETHOD
ENDCOMPONENT
```

# 4  COMMUNICATION AND SYNCHRONIZATION

Multimedia data streams are transmitted in arbitrarily structured flow graphs of interconnected components. Ensuring a satisfying stream quality over long periods of time while using current computer and network equipment makes the reservation of resources inevitable. Furthermore, due to the temporal dimension of time dependent data streams, there is a need to specify and control temporal properties of streams. Setting initial parameters like data rate or start values has to be enabled as well as scaling (i.e. changing speed or direction) at presentation time. The appropriate control interface in CINEMA is the media clock. However, an interface that only allows to handle individual data streams is insufficient. Due to tight relationships between different streams, they need to be grouped together and be handled as a unit. This facilitates the control over complex scenarios and is a prerequisite for specifying synchronization relationships between data streams. Especially, the latter is essential in a multimedia system where the quality of a presentation of time dependent data streams strongly depends on observing given synchronization requirements (e.g. lip synchronization of audio and video where the tolerable skew is in the range of 80 ms [13]). The grouping of data streams has to be supported by concepts that are adaptive to the dynamics of interactive and cooperative multimedia applications where at any time new users enter running applications (e.g. teleconferencing) and others leave. In CINEMA, the means to group control interfaces, to handle them as a unit and to specify synchronization relationships is given by the concept of clock hierarchies. In the following, the concepts to meet the requirements are explained in detail.

# 4.1 Session

In *CINEMA*, a session is the abstraction of resource reservation. It is associated with a set of quality of service parameters. By creating a session, a client causes the *CINEMA* system to reserve the resources that are needed to guarantee the specified quality of service requirements. This is done in an all-or-nothing fashion. After a session has been established, the transmission and processing of multimedia data may be started.

A session encompasses parts of the flow graph which is defined by a client. Its actual extension is defined by specifying a set of source and sink components. Intermediate components and interconnecting data paths are determined from the data flow graph by the *CINEMA* system. For example, a point-to-point audio session may be created by the following statement. It describes the components and their ports that are part of the session as well as the desired quality of service parameters:

```
create_session(micro  ->port("audio"),
               speaker->port("audio"),
               QoS(Rate(min = 8000, max = 44100),
                   SampleSize(min = 8, max = 16),
                   Delay(min = 50, max = 150)));
```

In *CINEMA*, quality of service is treated on different levels of abstraction. With sessions, application-specific quality of service specifications are associated. They represent the presentation quality a client wants to achieve at sinks. High-level quality of service parameters, such as picture size and picture rate, depend on the stream type. As resource reservation is independent from the application-level semantics of data streams, high-level parameters are mapped to low-level parameters, such as packet size and packet rate. Low-level parameters are based on parameters used in reservation protocols for multimedia transport systems (e.g. SRP [14], ST-II [15]).

The architecture of resource reservation is separated into two layers: global and local resource management. The global resource management is responsible for negotiation of the quality of service parameters at all the nodes participating in a session and mapping the high-level onto low-level parameters by using a distributed resource reservation protocol. Quality of service parameters are specified at sinks and transferred and negotiated in a sink-to-source direction. Our resource reservation protocol bases on ideas used in RSVP [16] and is designed to perform in arbitrary structured networks of components, which are distributed over any number of nodes (end-to-end reservation, for details see [17]). The local resource management reserves the resources as they are demanded by components or links. This leads to the implementation of several resource managers at each node, one for each individual resource (e.g. for memory, CPU utilization, network bandwidth). To perform resource management for CPUs, we have implemented a split-level scheduler using a modified rate-monotonic algorithm [18].

The success or failure of the establishment of a session determines whether a given application can be started and maintained according to the specified quality of service. Thus, creating a session is the prerequisite to transmit and process data units. Based on this, the following sections describe how temporal properties of streams are specified and data streams are controlled during run time.

## 4.2 Clocks

The temporal dimension of continuous media streams is defined by so-called media time systems. The media time system associated with a stream is the temporal framework to determine the media time of the stream's data units. In CIN-EMA, media time systems are provided by media clocks (or clocks for short). A clock $C$ is defined as follows: $C ::= ( R, M, T, S )$. The clock attributes have the following meaning: $R$ determines the ratio between media and real-time. $M$ is the start value of the clock in media time, i.e. the value of the clock at the first clock tick. $T$ is the start time of the clock in real-time. $S$ determines the speed of the clock. Media time progresses in normal speed if $S$ equals 1. A speed larger than 1 causes the clock to move faster, a speed smaller than 1 causes it to progress slower, and a negative speed causes it to move backwards. A clock relates media time to real-time. It "ticks" after it has been started and media time ($m$) can be derived from real-time ($t$): $m = M + S \cdot R (t - T)$

Clocks are the basic abstraction for clients to control the flow of media streams. They may be attached to components. Clocks attached to sink components control the temporal progress of data streams processed by those components. This is expressed more precisely by the **clock condition**: a data unit having media time $m$ is processed at real-time $t$ only if the controlling clock is ticking and its value equals $m$ at time $t$. Conceptually, this means that the presentation of a stream is started, paused or scaled when the controlling clock is started, halted or the clock speed is changed, respectively. Clocks attached to source components are typically required in flow graphs where multiple sources contribute data to a given sink (e.g. in a mixer scenario). In this case, source clocks are needed to individually start sources and to determine their start values. For more details on source clocks refer to [19].

The most important clock operations for controlling streams are the following. The operation Start(M) starts the clock at media time M, by doing this it starts the controlled stream as well. The clock attribute $T$ is set to the real-time at which the clock is actually started. Halt(M) halts the clock when it reaches clock value M, i.e. the stream controlled by this clock is paused. Prepare(M) prepares the starting of the clock at media time M by preloading the buffers along the communication paths of the controlled stream. After Prepare has been performed, the clock can be started immediately when Start is issued. Clear() clears the internal buffers associated with the controlled stream. Scale(M,S) changes the speed of the clock to S when media time M is reached, i.e. it scales the stream controlled by the clock.

In the simple scenario shown in Figure 5, clock $C$ controls the presentation of a video stream. The play-out is started with frame 15. The play-out rate is doubled when the presentation reaches frame 3000, and the presentation is halted when reaching frame 5000.

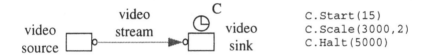

C.Start(15)
C.Scale(3000,2)
C.Halt(5000)

**Figure 5**   Controlling a Video Stream

## 4.3  Clock Hierarchies

In this section, we will introduce the notion of a clock hierarchy, which is the basic abstraction for grouping media streams, controlling groups of streams, and stream synchronization.

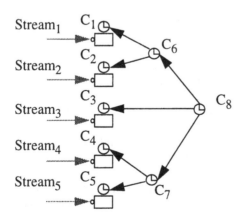

**Figure 6**   Grouping Streams

Remember that clocks attached to components control the streams processed by them. A number of streams can be grouped by linking their controlling clocks in a hierarchical fashion to a common clock, which then controls the entire group. Stream groups can be grouped again to groups at a higher level. In the example given in Figure 6, clock $C_6$ controls streams $S_1$ and $S_2$, while $C_7$ controls $S_4$ and $S_5$. $C_8$ controls the subgroups represented by $C_6$ and $C_7$ as well as stream $S_3$, and thus all streams in the given scenario can be started, halted or scaled collectively by means of this clock.

A clock operation issued at a clock not only affects this clock but the entire (sub)hierarchy of this clock. Conceptually, an operation called at a clock is **propagated** in a root-to-leaf direction through the clock's (sub)hierarchy, where it is performed at every clock in this hierarchy. In general, clock operations can be issued at every level of the clock hierarchy. Additionally, clock hierarchies may dynamically grow and shrink even if clocks are ticking. This feature together with the capability of halting and starting individual subhierarchies is very important in interactive applications, especially in those where multiple users with their individual needs participate in the same application.

Clocks provide individual media time systems which may relate to each other in various ways. Clock synchronization and propagation of clock operations is done on the basis of so-called **reference points**. A reference point defines the temporal relationship of two media time systems. More precisely, reference point $[C_1 : P_1, C_2 : P_2]$ defines that media time $P_1$ in $C_1$'s time system corresponds to media time $P_2$ in $C_2$'s time system, which means that $P_1$ and $P_2$ relate to the same point in real-time (see Figure 7). Given this reference point, media time can be transformed from one to the other time system as follows:

$$m_2 = (m_1 - P_1) \cdot \frac{S_2 R_2}{S_1 R_1} + P_2$$

Clocks may be linked in two different ways: a link may establish either a **control** or a **synchronization** relationship between two clocks. A control relationship between two clocks enables the propagation of clock operations without synchronizing them. Typically, control relationships are defined in settings where groups of streams are to be controlled collectively and a rather loose temporal coupling of the grouped streams is sufficient. Although **control hierarchies** include reference points, this information is considered only when clock operations are propagated to automatically transform the operation's arguments. However, after a hierarchy has been started, its clocks may drift out of synchronization and may be manipulated arbitrarily. For example, two different subhierarchies of the same hierarchy may be scaled in different ways, or clocks in the hierarchy may be halted and continued at any later time with arbitrary start values.

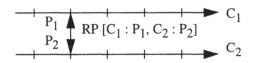

**Figure 7**  Transforming Media Time

A synchronization relationship goes a step further. In addition to propagation, it ensures that the clocks involved progress synchronously. From the clock condition introduced in the previous section it can be concluded that two streams are synchronized if their controlling clocks are synchronized. Thus, **synchronization hierarchies** are a general and very powerful concept to specify arbitrary synchronization requirements between media streams. The structure of the synchronization hierarchy specifies which streams have to be synchronized, while the reference points in the hierarchy define how the temporal dimensions of the streams relate to each other. The system guarantees that all streams controlled by the clocks of the hierarchy are processed synchronously. When a subhierarchy is halted and started once again at a later point in time, this is performed in conformance with the temporal constraints.

## *Example*

Figure 8 shows a simple telecooperation scenario with two users. Subject to the cooperation is an experiment shown on video $V_2$. We assume that additional speech channels exist which allow the users to talk to each other. The two users commonly view $V_2$. To ensure that both see the same information at the same time, $V_2$ must be played out synchronously. Besides $V_2$, user 1 views video $V_1$, which shows the same experiment from a different perspective. Consequently, $V_1$ and $V_2$ are to be synchronized. User 2 additionally views video $V_3$, which shows a similar experiment. Since the two experiments roughly correspond to each other in their temporal dimension, $V_1$ and $V_3$ are grouped by a control relationship. We assume that media time 500 in $V_3$ corresponds to media time 5 in $V_2$.

The presentation of all video streams can be started by issuing `Start` at clock $C_5$. Moreover, this clock can be used to collectively scale, pause and restart the entire configuration. User 1 may pause $V_1$ or $V_2$ by halting $C_1$ or $C_2$, respectively. Halted clocks may be continued in a synchronized fashion, i.e. after restart of $C_2$, for example, the presentation of $V_2$ is not only synchronized with $V_1$ but also with $V_2$'s presentation at the site of user 2. Since $C_3$ and $C_4$ are linked with a control edge, $V_3$ can be scaled, paused and restarted at any position independent of $V_1$'s and $V_2$'s state of the presentation. So, the presentation $V_3$ can be adjusted manually as needed.

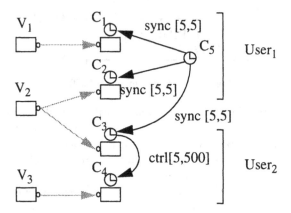

**Figure 8**   A Simple Telecooperation Scenario

If another user desires to join the scenario, the clock hierarchy has to be extended dynamically. After the corresponding session has been established, the clock controlling $V_2$'s presentation at the new user's site is linked by a sync edge to clock $C_5$. By issuing the start operation, $V_2$'s presentation is started synchronously to the ongoing presentations.

## 5   CLOCK HIERARCHIES AND NESTING

In the context of synchronization, nesting means that arbitrary complex clock hierarchies may be defined within compound components and thus remain invisible for the components' outside world. A clock hierarchy of a compound component is defined at the time the component is composed and specifies internal synchronization and control relationships between the clocks defined within this component. Only the root of internal clock hierarchies is exported and thus becomes visible to the components' outside world. The operations issued at an exported clock are propagated through the clock hierarchy and thereby control the internal processing. Exported clocks may again be involved in clock hierarchies at higher levels of abstraction.

The compound component shown in Figure 9 provides the abstraction of a television set, capable of playing out a video stream and two audio streams in a synchronized fashion. The component shown contains two basic components, a video decompression component ($D$) and sink component implementing a video output window ($W$). In addition, it includes another compound component, which consists of two filter components ($F$) and two speaker components ($S$). The nested compound component provides the abstraction of an audio output device, whose operation is controlled by clock $C_2$. The TV component exports clock $C_1$, which is used to start, pause or scale the audio-visual output.

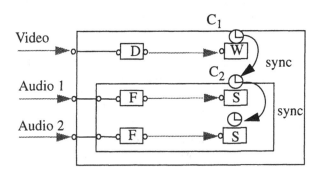

**Figure 9**   Nested Components

# 6   CONCLUSIONS

The efficient development of distributed multimedia applications requires abstractions and services that are provided by a specialized software layer. Such a middleware layer is based on general purpose operating systems and adds functions and protocols supporting distributed multimedia applications, including those for communication, synchronization and resource management. The *CINEMA* system is such a middleware layer. Our paper focused on the description of the service interface of *CINEMA*. We described components that provide multimedia processing functionality and may be nested to facilitate the reusability of software and to achieve higher levels of functional abstractions. Distributed multimedia applications are created by interconnecting the components' ports with links which allows the definition of arbitrary flow graphs. Before starting the flow of data units, the creation of sessions results in the reservation of system resources that are needed to ensure quality of service requirements. With media clocks and clock hierarchies we proposed abstractions to control individual data streams as well as groups of streams. We discussed the usage of clock hierarchies to specify synchronization relationships between data streams and showed how they may be used to handle the requirements of dynamic, interactive and cooperative multimedia applications. Finally, it was outlined how clock hierarchies are used to control the propagation of operations in compound components.

When implementing multimedia system services, certain requirements arise that have to be met by the operating systems *CINEMA* is layered on. For example the ability to schedule multiple independent multimedia processing tasks by observing real-time deadlines requires the support of real-time scheduling algorithms as well as preemptive threads, which the processing functions are mapped to. Current operating systems only partially fulfil this requirement. Furthermore, multimedia data is transmitted between components on the same node as well as on different nodes. The latter involves making use of transport protocols. When designing the CINEMA prototype that is based on IBM AIX and DCE as well as on the SUN Solaris operating system, it was decided to encapsulate multimedia transmission functionality into link objects. This offers two major advantages. It allows to use identical interfaces for data transmission between local and remote components which simplifies the configuration of applications significantly. Moreover, it decouples the CINEMA implementation from the transmission mechanism that actually is used. Due to the lack of a real-time transport system, in our current prototype link objects are based on UDP. For the next version it is planned to replace UDP by a real-time protocol which only requires a reimplementation of the link object. However, all other parts of the system are not affected.

The implementation of the CINEMA prototype is still in progress. The first version is working. It supports a restricted set of the functionality described in this paper. For example, it is possible to establish applications in a distributed environment and to control and to synchronize the flow of data units in limited configurations. Our future work is directed towards extending the prototype and gaining more experience in using our abstractions by experimenting with applications.

## REFERENCES

[1]   G. Coulson, G. S. Blair, P. Robin, D. Shepherd. Extending the Chorus Micro-Kernel to Support Continuous Media Applications. *4th International Workshop on Network and Operating Systems Support for Digital Audio and Video*, pp. 49–60, 11 1993.

[2]   M. Rozier, V. Abrossimov, F. Armand, I. Boule, M. Gien, M. Guillemont, F. Herrmann, C. Kaiser, S. Langlois, P. Léonard, W. Neuhauser. Overview of the Chorus Distributed Operating System. *Chorus Systémes CS/TR-90-25*, 4 1990.

[3]   J. Kramer, J. Magee. Dynamic Configuration for Distributed Systems. *IEEE Transaction on Software Engineering*, SE-11(4):424–436, 4 1985.

[4]   J. Magee, J. Kramer, M. Sloman, and N. Dulay. An Overview of the REX Software Architecture. *2nd IEEE Computer Society Workshop on Future Trends of Distributed Computing Systems*, 10 1990.

[5]   Apple Computer Inc., Cupertino, CA, USA. *QuickTime Developer's Guide*, 1991.

[6]   IBM Corporation. *Multimedia Presentation Manager Programming Reference and Programming Guide 1.0, IBM Form: S41G-2919-00 and S41G-2920-00*, 3 1992.

[7] D. P. Anderson, R. Govindan, G. Homsy. Abstractions for Continuous Media in a Network Window System. *Technical Report UCB/CSD 90/596, Computer Science Division, University of California, Berkeley*, 11 1990.

[8] R. B. Dannenberg, T. Neuendorffer, J. M. Newcomer, D. Rubine. Tactus: Toolkit-Level Support for Synchronized Interactive Multimedia. *3nd International Workshop on Network and Operating System Support for Digital Audio and Video*, 11 1992.

[9] Interactive Multimedia Association, Compatibility Project, Annapolis, MD, USA. *Request for Technology: Multimedia System Services, Version 2.0, available via ftp from ibminet.awdpa.ibm.com*, 11 1992.

[10] Hewlett-Packard Company, International Business Machines Corporation, SunSoft Inc. *Multimedia System Services, Version 1.0, available via ftp from ibminet.awdpa.ibm.com*, 7 1993.

[11] F. DeRemer, H. Kron. Programming-in-the-Large vs. Programming-in-the-Small. *Conference on Reliable Software*, pp. 114–121, 1975.

[12] R. G. Herrtwich. Time Capsules: An Abstraction for Access to Continuous-Media Data. *The Journal of Real-Time Systems, Kluwer Academic Publishers*, pp. 355–376, 3 1991.

[13] R. Steinmetz, C. Engler. Human Perception of Media Synchronization. *Technical Report 43.9310, IBM ENC, Heidelberg, Germany*, 1993.

[14] D. P. Anderson, R. G. Herrtwich, C. Schaefer. SRP: A Resource Reservation Protocol for Guaranteed-Performance Communication in the Internet. *Technical Report UCB/CSD 90/562, Computer Science Division, University of California, Berkeley*, 2 1990.

[15] C. Topolcic. Experimental Internet Stream Protocol, Version 2 (ST-II). *RFC 1190*, 10 1990.

[16] L. Zhang, S. Deering, D. Estrin, S. Shanker, D. Zappala. RSVP: A New Resource Reservation Protocol. *IEEE Network*, 9 1993.

[17] M. Häuptle. Development of a Resource Reservation Protocol for Distributed Multimedia Applications (in German). *Master's Thesis, University of Stuttgart/IPVR*, 4 1994.

[18] I. Barth. Extending the Rate Monotonic Scheduling Algorithm to Get Shorter Delays. *To be published: International Workshop on Advanced Teleservices and High-Speed Communication Architectures, Heidelberg, Germany*, 1994.

[19] K. Rothermel, T. Helbig. Clock Hierarchies: An Abstraction for Grouping and Controlling Media Streams. *Technical Report 2/94, University of Stuttgart/IPVR*, 4 1994.

# 16

# APPLICATION LAYER ISSUES FOR DIGITAL MOVIES IN HIGH-SPEED NETWORKS

## W. Effelsberg, B. Lamparter, R. Keller

*Praktische Informatik IV,*
*University of Mannheim, 68131 Mannheim, Germany*

## ABSTRACT

Digital movie system require a new generation of communication protocols. They are very demanding in terms of bandwidth, delay jitter, end-to-end delay, and multicast support within the network. We concentrate on application layer issues for digital move transmission. After introducing an application layer architecture, we present a novel, adaptive forward error correction scheme for movies, combining high correction probability with low overhead. We also present an efficient compression scheme for software movie systems, with better runtime properties than standardized movie compression algorithms.

## 1 INTRODUCTION

Networked digital movie systems require a completely new protocol stack. The conventional protocols were designed for low transmission speeds and relatively fast end systems; the applications were data-oriented. Today the paradigm has changed to high transmission speeds over fiber optics cables, and the inclusion of continuous media, in particular digital audio and video, for multimedia applications.

A protocol stack for this environment must have new protocols in *all* layers guaranteeing

- high bandwidth (at least 8 MBit/s per channel)
- low delay jitter

- low end-to-end delay

- efficient multicast.

Recent research has concentrated on the lower layers only, and indeed the above characteristics are increasingly met by the new generation of protocols for layers 1 to 4. In particular, the ATM architecture and protocols provide appropriate layers 1 and 2, and network and transport protocols such as XTP, ST-II and OSI-HSTP solve most of the problems with error recovery, flow control, guaranteed quality of service and multicast for layers 3 and 4. However very little work was reported on upper layer architecture and protocols for the new paradigm. These are the topic of our paper.

We discuss our work in the context of XMovie, a software movie system for Unix workstations. The system is implemented and running, and we can use it as a testbed for our application layer algorithms. This allows us to report not only concepts, but also experimental results.

## 2 THE XMOVIE CLIENT/SERVER ARCHITECTURE

The XMovie system is an experimental prototype developed at the University of Mannheim [8]. It can transmit stored movies over high–speed networks and display them in windows of the X window system. Unlike other digital movie systems, it requires no special hardware in workstations. Relying on special hardware seems to be too constraining for innovative multimedia systems, even though it can help provide better performance in the short term. Our approach is to use software solutions wherever possible, and special hardware only when it is absolutely necessary (e.g. a video digitizer board as the source of a live movie).

The system has been operational for about two years and was ported to DEC-station 5000, to Sun SPARCstation 10 and to IBM RS/6000. The movies are currently transmitted over Ethernet or FDDI, using IP and UDP and our own Movie Transmission Protocol [10] on top of these. The movies we use for experimentation are digitized videos and computer–generated films.

Three alternative display architectures for movies on workstations have been reported in previous work: The first one uses a blue–box architecture like

Pandora's box [5]. The special hardware needed limits flexibility and makes integration with other windows difficult; it also limits the number of movie windows. The second alternative shows a movie as a fast series of single X Window images, usually implemented using shared memory. In this approach the X client has to send a request for each image to the X server. The MPEG player of UC Berkeley uses this approach [15]. Its main disadvantages are that synchronization is up to the application, that the X client and X server must be co–located to gain highest speed, and that multiple movies create considerable overhead. The third alternative is the XMovie approach, an extension to the window system, with a Movie Transmission Protocol between movie server and movie client.

We have extended the X server to process several movie orders from X clients simultaneously. The X server reads the frames sent by a movie server directly from the net and displays them in a window (see Fig. 1).

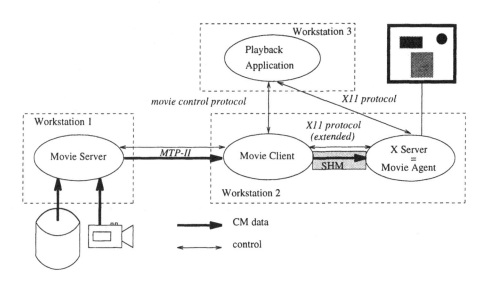

**Figure 1**   XMovie System Architecture

The main advantage of our approach is that the X server can schedule the display of images according to its own requirements, i.e. without deterioration of other X clients, and to those of the movie, i.e. frame rate and isochronous flow. If the X server is heavily loaded it finds a compromise between the two.

The system has been operational for about two years and was ported to DEC-station 5000, to Sun SPARCstation 10 and to IBM RS/6000. The movies are currently transmitted over FDDI, using IP and UDP and our own Movie Transmission Protocol MTP [10 ] on top of these. The movies we use for experimentation are digitized videos and computer-generated films.

In this paper we address two important issues in the context of our application layer for movies: First, we present the novel forward error correction scheme AdFEC integrated into our Movie Transmission Protocol. We believe that this is the place in the protocol stack where error correction should be done, for the following reason: The new generation of protocols supports a wide variety of different applications, including connection-oriented data transfer, connection-less data transfer, acknowledged datagrams (sometimes called "transactions") and continuous streams with guaranteed quality of service. All of these have quite different error correction requirements. Lower layer protocols which are common to all applications cannot support specific requirements efficiently.

Second, we introduce an improved compression scheme for digital movies called XCCC. Research prototypes usually do not include the compression/decompression algorithm into the stream transfer layer. This usually leads to additional copy operations between the compression module and the application layer of the communication subsystem. Such copy operations are harmful in high-speed networks where the bottleneck is the CPU. Our approach of integrating the compression algorithm with the movie application layer saves at least one copy operation.

# 3   AN ADAPTABLE FORWARD ERROR CORRECTION SCHEME FOR DIGITAL MOVIES

The error handling method in traditional communication protocols is error detection and retransmission. This method is inappropriate for distributed multimedia systems for two reasons: It introduces variable delay unacceptable for isochronous streams, and it is very inefficient and difficult to use in the multicast environment typical for many multimedia applications. We propose AdFEC, an Adaptable Forward Error Correction scheme based on binary polynomial algebra. It produces an adaptable amount of redundancy allowing different packet types to be protected according to their importance. The scheme was implemented in the framework of the XMovie project and proved to be very efficient.

Single bit errors rarely occur in modern networks, especially if these are based on fiber optics. The main source of errors is packet loss in the switches. Current FEC procedures that focus on the correction of bit errors do not solve this problem.

The well-known error correcting codes serve primarily to correct individual bit errors. Very few articles address the problem of reconstructing lost packets [18, 13, 1, 14]. These articles deal with packet loss in ATM networks. All packets in the data stream are protected by means of the same method, and with the same redundancy.

The AdFEC scheme is capable of assigning different priorities to different parts of the data stream. The amount of redundancy for FEC is chosen according to the priority. A digital data stream for a movie or for audio contains more than just the digitized image/audio contents. It also contains information that must not be lost under any circumstances, such as control instructions, format data, or changes in the color lookup table. Typically a higher error rate can be tolerated for content parts than for control parts, but all packets have to arrive on time. For example we can assign priorities as follows:

**Priority 1:** Segments that may not be lost under any circumstances (e.g. control and format information as well changes in the color lookup table)

**Priority 2:** Segments whose loss clearly adversely affects quality (e.g. audio)

**Priority 3:** Segments whose loss is slightly damaging (e.g. pixel data in a video data stream)

For none of the three priorities is retransmission a tenable option. Starting from an already low rate of loss, third-priority packets can be transmitted without protection, second-priority packets should be protected by means of FEC with minimum redundancy, and first-priority packets by means of FEC with high redundancy.

## 3.1  Creating Redundancy for Forward Error Correction

Traditional error-correcting codes (e. g. Reed-Solomon) were designed for the detection and correction of bit errors [12]. Since the demand now exists to also restore entire packets, new codes must be found. Specifically, errors need no longer be located, the lost bits are known. A feature of traditional error-correcting codes is their ability to locate damaged bits. The price of this feature is a great deal of redundancy. At issue here is to devise a code that restores the lost packets at a known error location.

**Example 1**: Two packets $p_1$ and $p_2$ are to be sent. A redundancy of 100% is taken into account, i. e. two additional packets may be generated. These additional packets are sent together with the original packets. In the event of packet loss, the original packets $p_1$ and $p_2$ must be restored from the remaining packets (see Figure 2). In this case two operations (labeled ∘ and •) are necessary, with whose help the redundant packets can be generated.

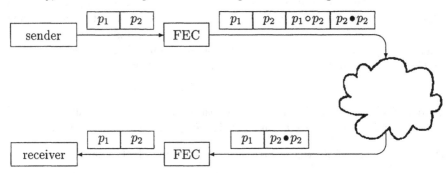

**Figure 2**  Principle of forward error correction to prevent packet losses (assume the network loses the packets $p_1$ and $p_1 \circ p_2$)

We can now define the problems in mathematical terms.

**Definition:** Bit sequence $\mathbf{B}_l = \{0, 1\}^l$, for fixed $l \in \mathbf{N}$.

**given:** $n$ packets $P = \{p_1, p_2, \ldots, p_n \in \mathbf{B}_l\}$

**find:** $n + m$ packets $Q = \{q_1, q_2, \ldots, q_{n+m} \in \mathbf{B}_l\}$ such that upon the arrival of at least $n$ packets out of $Q$ all packets of the set $P$ can be restored.

**Example 2**: $n = 2, m = 1$:

Choose $q_1 = p_1, q_2 = p_2, q_3 = p_1 \oplus p_2$

$\oplus$ is the operator for binary exclusive or (XOR). If, for example, $q_1$ is lost, $p_1$ can be restored with $p_1 = q_2 \oplus q_3$.

A total of 16 binary operators can be defined combining two bits, namely all possible combinations of zeroes and ones in a four-row value table. In order to construct our packets for $Q$, operators are required whose result can be used for the unambiguous reconstruction of any bit fields. The only binary operator suitable for this is the XOR-operator and its negation, the equivalence. Therefore on the basis of just the 1-bit operators only one redundancy packet can be generated, all other packets would not allow an unambiguous reconstruction in all cases.

In order to generate additional independent packets, a field containing $2^n$ elements ($n > 1$) must be sought. Such fields can be generated with the aid of polynomial algebra [12]. Table 1 shows the operations + and * for a field containing four elements.

| + | 00 | 01 | 10 | 11 | | * | 00 | 01 | 10 | 11 |
|----|----|----|----|----|---|----|----|----|----|----|
| 00 | 00 | 01 | 10 | 11 | | 00 | 00 | 00 | 00 | 00 |
| 01 | 01 | 00 | 11 | 10 | | 01 | 00 | 01 | 10 | 11 |
| 10 | 10 | 11 | 00 | 01 | | 10 | 00 | 10 | 11 | 01 |
| 11 | 11 | 10 | 01 | 00 | | 11 | 00 | 11 | 01 | 10 |

**Table 1**  The operations + and * for a field containing four elements

For $n = m = 2$ the operators $\circ$ and $\bullet$ can, for example, be defined as follows (the field is labeled GF for Galois Field):

$$a, b \in \mathrm{GF}(2^2) \Rightarrow \quad a \circ b = a + b$$
$$a \bullet b = a + 10 * b$$

The two redundancy packets are thus defined as

$$c = a + b$$

and

$$d = a + 10 * b$$

In the case of loss of two data packets, the calculations in Table 2 are to be carried out.

| arriving Packets | Calculations for reconstructing $a$ | Calculations for reconstructing $b$ |
|---|---|---|
| $a, b$ | - | - |
| $a, c$ | - | $a + c$ |
| $a, d$ | - | $11 * (a + d)$ |
| $b, c$ | $b + c$ | - |
| $b, d$ | $d + 10 * b$ | - |
| $c, d$ | $11 * c + 10 * d$ | $10 * (c + d)$ |

**Table 2** Calculations for reconstruction of lost packets

Returning to our notation above we set $p_1 = a, p_2 = b, q_1 = c$ and $q_2 = d$. Thus, if any two packets are lost, the contents of $p_1$ and $p_2$ can be reconstructed from the arriving packets.

Generally speaking, fields containing $p^n$ elements ($p$ is prime, $n > 0$) can be derived. Of particular interest to the computer scientist are of course those with $p = 2$. For this reason only such fields will be considered in the following. The book by Lin and Costello [12] provides in its appendix a list of the possible generator polynomials for $GF(2^n)$ to $n = 10$.

## The Power of Polynomial Algebra for Forward Error Correction

The fields constructed in the preceding section are to be examined regarding their power for generating redundant packets. Of particular interest is the number of *independent* packets that can be generated. Given the two packets $a$ and $b$, generation occurs in the form $\lambda a + \mu b$, with $\lambda, \mu, a, b \in GF(2^n)$.

Thus, given a field $GF(2^n)$ and a number of packets $a_1, \ldots, a_k$, the question is: How many independent packets $l$ can be generated by the given field from $k$ packets? The more packets generated and transmitted, the greater the probability that the recipient will be able to reconstruct the packets sent from those that arrived. It has already been explained above that by means of a field $GF(2)$ only one single additional packet can be generated, independent of the number of given packets. The extension to the field GF(2) apparently is of little use in this respect. For this reason the fields $GF(2^n)$ are examined.

Generally the contents of the output packets for two input packets $a$ and $b$ adheres to the following scheme:

Output Packets:

$p_{ij} = \lambda * a + \mu * b$ with $i, j = 1 \ldots 2^n$ and $\lambda, \mu \in GF(2^n)$ (see Table 3)

| | | | | | | | |
|---|---|---|---|---|---|---|---|
| 1. | $0 +$ | $0$ | | 9. | $\alpha a$ | $+$ | $0$ |
| 2. | $0 +$ | $b$ | | 10. | $\alpha a$ | $+$ | $b$ |
| 3. | $0 +$ | $\alpha b$ | | 11. | $\alpha a$ | $+$ | $\alpha b$ |
| 4. | $0 +$ | $(1+\alpha)b$ | | 12. | $\alpha a$ | $+$ | $(1+\alpha)b$ |
| 5. | $a +$ | $0$ | | 13. | $(1+\alpha)a +$ | | $0$ |
| 6. | $a +$ | $b$ | | 14. | $(1+\alpha)a +$ | | $b$ |
| 7. | $a +$ | $\alpha b$ | | 15. | $(1+\alpha)a +$ | | $\alpha b$ |
| 8. | $a +$ | $(1+\alpha)b$ | | 16. | $(1+\alpha)a +$ | | $(1+\alpha)b$ |

**Table 3**  Generateable packets in $GF(2^2)$

Some of the packets generated in this manner are linearly dependent, and their transmission thus affords no advantage for reconstruction. The transmission of the first packet is obviously useless because it is always equal to zero. The pairs of linearly dependent packets can now be combined into classes. $n - 1$ elements belong to every class, in the example three elements of the table. The result yielded here is the set $\{\{b, \alpha b, (1+\alpha)b\}, \{a+b, \alpha a + \alpha b, (1+\alpha)a + (1+\alpha)b\}, \{a + \alpha b, \alpha a + (1+\alpha)b, (1+\alpha)a + b\}, \{a + (1+\alpha)b, \alpha a + b, (1+\alpha)a + \alpha)b\}, \{a, \alpha a, (1+\alpha)a\}\}$.

A more detailed discussion of the systematic construction of linearly independent packets, with several theorems and algorithms, can be found in [9].

Let us now come to the corrective power and the overhead of our scheme. The more linearly independent packets we include in the transmission, the higher the corrective power (i.e. chance of restoration), but the higher also the overhead. For example, only minimal protection is provided by a single additional packet for every ten data packets. This results in a redundancy of only 10 percent. But with our scheme it is now easy to generate a larger number of redundant packets, thereby strongly increasing the reconstruction probability. For example, using a code out of the field $GF(2^8)$, 257 pairs of linearly independent packets could be generated out of only two data packets. This goes to the other extreme in increasing the data rate, but providing much better correction probability.

As we have seen, the method described enables the generation of an *adaptable redundancy*. Using the same method of calculation at the sender and receiver, the different segments of the data stream can be protected with varying degrees of correction probability and overhead. Therefore we call our method AdFEC (Adaptable Forward Error Correction).

## 3.2   The Corrective Power of AdFEC

In this section we analyze the corrective power of the AdFEC method. We compare AdFEC with two alternatives: duplicating each packet, or no redundancy at all (i. e. no error correction).

Let n be the number of packets to be send and m be the number of packets that are added for error correction. Furthermore, let q be the probability that a packet arrives (i. e. $1 - q$ is the probability of loosing a particular packet). We assume that the loss of packets is an independent process. We use the following notation to denote the probabilities that the original n packets can be recovered:

$p(n, m)$     under the assumption that m additional packets are added;
$p_d(n)$      under the assumption that every packet is send twice;
$p_n(n)$      no forward error correction.

To obtain a formula for $p(n, m)$ note that the probability that exactly i out of the $n + m$ packets get lost is $q^{n+m-i} (1 - q)^i \binom{n + m}{i}$. In order to reconstruct the original n packets at most m packet losses are tolerable. Taking the sum over all possible cases we get

$$p(n, m) = \sum_{i=0}^{m} q^{n+m-i} (1 - q)^i \binom{n + m}{i}$$

For the analysis of $p_d(n)$ note that the original n packets can be reconstructed only in the case of least one of the two duplicates arrives. Thus, if i packets are lost, the original n packets arrived at the receiver if the lost packets were all different. There are $2^i \binom{n}{i}$ different cases for loosing exactly i different packets. This yields

$$p_d(n) = \sum_{i=0}^{n} q^{2n-i} (1-q)^i 2^i \binom{n}{i}$$

Clearly for the function $p_n(n)$ we have

$$p_n(n) = q^n$$

The overhead in the case of sending every packet twice is 100%. In the case where m additional packets are send the overhead is $100m/n\%$.

The following tables characterize the situation for $q = 0.9$ and $q = 0.7$. In the first case we have $p_n(10) = 0.35$ and $p_d(10) = 0.90$ and in the second case we have $p_n(10) = 0.03$ and $p_d(10) = 0.39$.

Probability $q = 0.9$

| numb. of add. packets | $p(10, m)$ | overhead in % |
|---|---|---|
| 1 | 0.697 | 10 |
| 2 | 0.889 | 20 |
| 3 | 0.966 | 30 |
| 4 | 0.991 | 40 |
| 5 | 0.998 | 50 |

Probability $q = 0.7$

| numb. of add. packets | $p(10, m)$ | overhead in % |
|---|---|---|
| 1 | 0.113 | 10 |
| 2 | 0.253 | 20 |
| 3 | 0.421 | 30 |
| 4 | 0.584 | 40 |
| 5 | 0.722 | 50 |

Thus, in the case $q = 0.9$ with an overhead of only 20%, nearly the same arrival probability can be achieved by our algorithm as in the case where every packet is sent twice (i. e. the overhead is 100%). In the case $q = 0.7$ the situation is similar: with an overhead of only 30% a better result can be achieved than with duplication. Moreover, the arrival probability with only 10% overhead is 69% (11%) compared to 35% (3%) without error correction. *Hence, our adaptive method is superior to the simple method of duplicating the packets, and with very little overhead it is possible to improve the arrival probability by a factor of 5 – 10 over no error correction.*

## 3.3   Performance of AdFEC

We have implemented the AdFEC algorithms described above in the XMovie system. We decided to use 0, 1 or 2 redundant packets for two packets, depending on the importance of the packets transmitted:

**Priority 3 ("loss does no harm"):** Packets are transmitted without redundancy.

**Priority 2 ("small amount of loss acceptable"):** Packets are transmitted in such a manner that for every two data packets A and B a third AdFEC packet A∘B is generated.

**Priority 1 ("loss not acceptable"):** Packets are transmitted in such a manner that for every two data packets A and B, two AdFEC packets A∘B and A•B are generated and transmitted.

In the framework of XMovie's MTP protocol, AdFEC protects the color lookup table transmission; damage to or a loss of the color lookup table within a movie data stream would have catastrophic results. In our implementation the field $GF(2^2)$ was used. Since the size of the color lookup table is only about 1 kbyte per image, compared to approx. 50 kbytes of pixel data, and a high degree of security is required, priority 1 was chosen was chosen for the color lookup table packets. Because reconstruction requires only XOR and multiplication by $\alpha$, AdFEC could be implemented with very few machine instructions. The multiplication uses a small table in memory: table lookup is more efficient than explicit computation at runtime, and can be carried out in just a few machine instructions. Since the addition corresponds to the XOR operation which is carried out in hardware, the total efficiency of AdFEC is very high. AdFEC was written in C on UNIX workstations (SUN-10 and DEC-5000). Because standard C was used exclusively, porting of the error correction procedure to other architectures is very easy.

# 4 EFFICIENT MOVIE COMPRESSION WITH XCCC

We now come to our second application layer algorithm, efficient compression for digital movies. The standardized compression techniques JPEG for still images [19] and MPEG for motion pictures [4] both include a Discrete Cosine Transform (DCT). This is a complex and computationally very demanding mathematical function. As a consequence, software motion pictures based on JPEG or MPEG are very slow, even on the most powerful CPUs available today, and it is generally assumed that these compression schemes will only work well with special hardware. However, special hardware makes a movie system much less flexible and less portable. It is thus desirable to develop alternative compression/decompression algorithms for movies which are optimized for computation in software on general purpose CPUs.

We propose an extension to the Color Cell Compression scheme for use in multimedia workstations. After a short introduction into Block Truncation Coding for monochrome images and Color Cell Compression for color images, we describe our eXtended Color Cell Compression (XCCC) scheme in detail. We have implemented XCCC and present experimental results on runtime performance and compression ratios.

## 4.1 Block Truncation Coding (BTC) and Color Cell Compression (CCC)

Our eXtended Color Cell Compression (XCCC) algorithm belongs to the family of block compression algorithms. Earlier examples from this family include Block Truncation Coding (BTC) and Color Cell Compression (CCC), and we begin with a brief introduction into these.

The Block Truncation Coding Algorithm [3] is used in the compression of monochrome images. The algorithm decomposes the whole image into blocks of size $n \times m$ pixels. These blocks are coded with two grey values $a$ and $b$ and a bit array. The bits in the bit array show which of the values $a$ or $b$ is to be used during decompression. For example, a 1 indicates that value $a$ should be used, a 0 stands for $b$.

If the original image used one byte per pixel, the storage requirement was 128 bits for a $4 \times 4$ image block. Storage of the compressed block requires 16 bits for the bit array plus one byte for each of the values $a$ und $b$. Hence BTC reduces storage from eight bits to two bits per pixel.

This basic version of BTC can be improved with a number of tricks [16]. In [17] a hierarchical encoding method for gray scale images is described.

If BTC is to be used for color images rather than for gray scales, the components (red, green, and blue, resp. chrominance and luminance) may be compressed separately. However, the CCC method then yields a much better compression rate [2].

Similar to BTC, the image is divided into blocks called "color cells". The two values $a$ and $b$ are now indices into a color lookup table (CLUT). The decompression algorithm works analogous to the BTC method.

The two values $a$ und $b$ can each be stored in one byte if the CLUT has 256 entries. Hence, the storage needed for a block of size $4 \times 4$ is two bits per pixel as with BTC (to be more exact, we would have to add the storage needed by the CLUT ($256 \times 3$ Bytes for the full image)). Color Cell Compression is not only one of the best compression algorithms, it is also one of the fastest [16].

As with the BTC algorithm, a number of possible improvements exist here as well:

- If the two colors $a$ and $b$ are nearly equal, or one color dominates in frequency of occurrence, only one color is stored, and no bit array is needed.

- If an image contains large areas with only small differences in color, those areas may be encoded with larger blocks.

- For movies cuboids may be used, with time being the third dimension, if the changes from frame to frame are small enough.

Our algorithm XCCC is based on the second idea. It uses $4 \times 4$, $8 \times 8$, and $16 \times 16$ blocks in a hierarchical fashion. Larger blocks tested in an earlier version yielded no further improvement.

## 4.2 Extensions to CCC

Our scheme extends CCC in two steps in order to improve compression ratio and runtime performance. We thus call it XCCC (extended CCC).

### *First step: Adaptive Block Sizes*

In many cases an image has large areas with small differences in the colors (i. e. in the background). These areas can be coded with fewer bits. We first investigate the optional use of larger rectangles. If an image contains large areas with few colors, these areas can be compressed with larger rectangles.

With XCCC the images are first decomposed into large blocks ($16 \times 16$) and, if necessary, these blocks are then subdivided. The algorithm for each block $B$ is:

1. Calculate the CCC coding of the block

2. If the actual block has the minimal block size, then Done.

3. Calculate the mean difference $\Delta e$ of the original pixel values and the values coded with CCC: $\Delta e = \sum_B |p - p'|_2$, where $p$ is the pixel value, $p'$ is the value of the same pixel after decompression.

4. If $\Delta e$ is smaller than a given constant, then Done.

5. Divide the block into four subblocks and use the algorithm recursively for each of these blocks.

The data for the simple CCC could be arranged in the data stream without any structuring information. But the output of the extended algorithm is a tree of blocks for each block. Hence, we have to store a complex data structure. This is done by adding a tag for each block. Figure 3 shows an example of the coding of an XCCC block. The tag is the logarithm of the length of the edge of the coded block. Blocks of minimal size need only one tag for four blocks because one minimal block is always followed by three more. After the tag we store the indices into the CLUT ($a$ and $b$) and then the bit array. If the length or width of the image is not divisible by 16, we divide the residual rectangle into $4 \times 4$ blocks and encode some of these subblocks as rectangles.

While the use of adaptive block sizes introduces a small additional overhead for compression, it results in much more efficient decompression for most images.

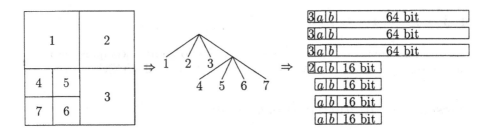

**Figure 3**   Coding tree of the XCCC algorithm

## Second step: Single color and color reuse

If an image has a large area with only one color, it is not necessary to store the bit array at all. XCCC does not store a bit array if the two color indices $a$ and $b$ are equal. There are two ways to let the decoder know that there is no bit array: First the encoder can store the two colors and the decoder will know from the equality that there will be no bit array. Second, the encoder can use a special tag and store only one color. XCCC uses the second method.

If we implement bit array suppression for single color blocks, larger squares will not always improve the compression ratio. Instead of a $32 \times 32$ square, XCCC uses four $16 \times 16$ squares. But some of these squares will have no bit array and hence the total compression ratio may be better. Though this can also happen with smaller blocks, our experience shows that this is rarely the case.

Colors in the neighborhood are often equal in images. Because of this fact, we can sometimes reuse colors from the block coded previously. Color reuse is also stored in the tag.

Further improvements are possible, and are described in detail in [11].

## 4.3   Experience: Compression Ratio and Decompression Speed

The main goal of the XCCC scheme is to allow rapid decompression with software decoders. Table 4 shows the decompression speed in images per second. The measurements of MPEG [4] were done with the MPEG player of the University of California at Berkeley [15]. We used three movies: The first and second movie (butterfly) is a raytraced sequence of 350 frames sized $320 \times 256$

and 780 × 576 resp. The third movie was digitized from an analog video of the University of Mannheim. It consists of scenes of buildings of the university (a castle) and other scenes depicting university life. Due to the analog origin, this movie consists of many different colors and color shadings. It is 320 × 256 in size and has 2000 frames.

The decompression was done on a DEC/alpha workstation with a 133Mhz CPU. The speeds are the real speeds viewed on the screen. The display adapter uses a color lookup table with 8 bits per pixel. XCCC uses the same technique internally thus requiring no conversion. In contrast, MPEG uses full color internally. Hence, the MPEG player has to dither in real-time. The player has several built in dithering methods. For the tests we used the fastest color dithering available ("ordered dithering").

| Movie | Size (pixels$^2$) | MPEG (frames/s) | XCCC (frames/s) |
|---|---|---|---|
| Butterfly | 320 × 240 | 10.5 | 60 |
| Butterfly | 780 × 576 | 2.1 | 6.8 |
| Castle | 320 × 240 | 7.8 | 24 |

**Table 4** Decompression speed of software decoders (in frames/s)

| Movie | Size | MPEG | JPEG | XCCC |
|---|---|---|---|---|
| Butterfly | 320 × 240 | 0.80%≙0.19bpp | 2.49%≙0.60bpp | 3.0%≙0.72bpp |
| Butterfly | 780 × 576 | 0.53%≙0.13bpp | 1.54%≙0.37bpp | 1.83%≙0.44bpp |
| Castle | 320 × 240 | 1.5%≙0.36bpp | 5.9%≙1.42bpp | 6.3%≙1.51bpp |

**Table 5** Compression ratios (compressed/full color size and bits per pixel)

The compression speed was measured on a DEC5000/133. For the small butterfly movie we got about 6.5 seconds per image with MPEG. Before XCCC can be started, the color lookup tables have to be computed. This step takes about 7 seconds per image. XCCC then takes about 2 seconds per image. Our compressor uses a total of about 10 seconds per image.

Our experiments show that XCCC can decompress images very fast. The quality of the XCCC compressed images is comparable to that of MPEG-compressed images. The advantage of MPEG is the size of the compressed movie. The size of XCCC compressed movies is about three to four times larger than that of MPEG movies. Hence, the domain of XCCC are local area networks with color workstations using the color lookup table technique. In this environment XCCC performs significantly better than MPEG.

## 5  CONCLUSIONS

We have briefly introduced the architecture of XMovie, an experimental movie system for high-speed networks. It can be used as a testbed for experimentation with application layer protocols for digital movies. There is growing interest in research and industry in such movies for use in video-on-demand systems, multimedia document retrieval, distance learning, distributed multimedia kiosk systems, and many other applications.

Whereas traditional protocol stacks handle error correction in the lower layers, modern high-speed networks have to support such a variety of different applications with different error correction requirements that only dedicated algorithms in the application layer are appropriate. Our AdFEC algorithm allows the creation of an adaptable amount of redundancy for packets within a data stream. An evaluation of AdFEC's corrective power shows its clear superiority over packet duplication. An implementation and experiments have shown that AdFEC can be executed very efficiently in software.

We have also presented XCCC, an algorithm to compress and decompress digital movies on standard color workstations at a reasonable speed without special hardware. We have shown that our algorithm is much faster than MPEG when implemented in software. On the other hand, MPEG gives a better compression ratio than XCCC.

## REFERENCES

[1] E.W. Biersack, "Performance Evaluation of Forward Error Correction in ATM Networks, Computer Communication Review, 22(4), October 1992, pp. 248–257

[2] G. Campbell, T. A. DeFanti, J. Frederikson, S. A. Joyce, A. L. Lawrence, J. A. Lindberg, and D. J. Sandin, "Two bit/pixel full color encoding, Computer Graphics, 1986.

[3] E. J. Delp and O. R. Mitchell, "Image Compression using Block Truncation Coding, IEEE Transactions on Communications, 1979.

[4] D. Le Gall, "MPEG: A Video Compression Standard for Multimedia Applications, Communications of the ACM, 34(4), 1991, pp. 46–58.

[5] A. Hopper, "Pandora – an Experimental System for Multimedia Applications, Operating Systems, 24(2), April 1990.

[6] B. Hornfeck, "Algebra, de Gruyter, Berlin, 1976.

[7] D.A. Huffman, "A method for the construction of minimum reduncancy codes, Proceedings IRE, 40, 1962, pp. 1098–1101

[8] B. Lamparter, W. Effelsberg, "XMovie: Transmission and Presentation of Digital Movies under X, Proc. 2nd International Workshop on Network and Operating System Support for Digital Audio and Video, Heidelberg, November 1991, LNCS 614, Springer-Verlag, Berlin Heidelberg, 1992, pp 328-339

[9] B. Lamparter, W. Effelsberg, Otto Böhrer, V. Turau, "Adaptable Forward Error Correction for Multimedia Data Streams, Technical Report TR-93-009, University of Mannheim, 1993, http://www.informatik.uni-mannheim.de/

[10] B. Lamparter, W. Effelsberg, N. Michl, "MTP: A Movie Transmission Protocol for Multimedia Applications, In: Multimedia92, 4th IEEE ComSoc International Workshop on Multimedia Communications, Monterey, California, April 1992, pp. 260–270. Extended abstract in ACM Computer Communications Review, Vol. 22, Juli 1992, pp. 71–72

[11] B. Lamparter, W. Effelsberg, "Extended Color Cell Compression – A Runtime-efficient Compression Scheme for Software Video, Proc. 2nd International Workshop on Advanced Teleservices and High-Speed Communication Architectures (IWACA-2), Heidelberg, 1994, Springer LNCS, to appear.

[12] S. Lin, J. Costello, "Error Control Coding, Prentice Hall, 1983.

[13] A. J. McAuley, "Reliable Broadband Communication using a Burst Erasure Correcting Code, Computer Communication Review, 20(4), September 1990, pp. 297–306

[14] H. Ohta and T. Kitami, "A Cell Loss Recovery Method Using FEC in ATM Networks, IEEE Journal on Selected Areas in Communications, 9(9), Dezember 1991, pp. 1471–1483

[15] K. Patel, B. C. Smith, L. A. Rowe, "Performance of a Software MPEG Video Decoder, In P. Venkat Rangan, editor, Proceedings of ACM Multimedia 93, Addison-Wesley, Aug 1993, pp. 75–82.

[16] M. Pins, "Analysis and choice of algorithms for data compression with special emphasis on images and movies (In German), PhD Thesis, University of Karlsruhe, Germany, 1990.

[17] J. U. Roy and N. M. Nasrabadi, "Hierarchical block truncation coding. Optical Engineering, 30(5), May 1991, pp. 551–556

[18] N. Shacham, P. McKenny, "Packet Recovery in High-Speed Networks using Coding, In: Proc. INFOCOM 90, San Francisco, Juni 1990.

[19] G. K. Wallace, "The JPEG Still Picture Compression Standard, Communications of the ACM, 34(4), April 1991, pp. 31–44.